Universes and Multiverses:

From a New Standard Model to a Physical Multiverse; The Big Bang; Our Sister Universe's Wormhole; Origin of the Cosmological Constant, Spatial Asymmetry of the Universe, and its Web of Galaxies; A Baryonic Field between Universes and Particles; Flatverse Extended Wheeler-DeWitt Equation

By the Power of Truth, I, while Living, have Conquered the Universe.

Faust - Goethe

Universes and Multiverses:

From a New Standard Model to a Physical Multiverse; The Big Bang; Our Sister Universe's Wormhole; Origin of the Cosmological Constant, Spatial Asymmetry of the Universe, and its Web of Galaxies; A Baryonic Field between Universes and Particles; Flatverse Extended Wheeler-DeWitt Equation

STEPHEN BLAHA, PH. D.

BLAHA RESEARCH

Blaha Research, P. O. Box 368, Auburn, NH 03032 USA

ISBN: 978-0-9893826-2-5

Cover Credits

Symbolic representation of universes connected by a wormhole. Cover design by Stephen Blaha © Copyright 2013. All Rights reserved.

Rev. 00/00/01 January 27, 2014

To My Wife Margaret

T

Some Other Books by Stephen Blaha

From Asynchronous Logic to The Standard Model to Superflight to the Stars (Blaha Research, Auburn, NH, 2011)

From Asynchronous Logic to The Standard Model to Superflight to the Stars; Volume 2: Superluminal CP and CPT, U(4) Complex General Relativity and The Standard Model, Complex Vierbein General Relativity, Kinetic Theory, Thermodynamics (Blaha Research, Auburn, NH, 2012)

The Algebra of Thought & Reality: The Mathematical Basis for Plato's Theory of Ideas, and Reality Extended to Include A Priori Observers and Space-Time; Second Edition (Pingree-Hill Publishing, Auburn, NH, 2009)

Quantum Theory of the Third Kind: A New Type of Divergence-free Quantum Field Theory Supporting a Unified Standard Model of Elementary Particles and Quantum Gravity based on a New Method in the Calculus of Variations (Pingree-Hill Publishing, Auburn, NH, 2005)

Quantum Big Bang Cosmology: Complex Space-time General Relativity, Quantum Coordinates, Dodecahedral Universe, Inflation, and New Spin 0, ½, 1 & 2 Tachyons & Imagyons™ (Pingree-Hill Publishing, Auburn, NH, 2004)

SuperCivilizations: Civilizations as Superorganisms (McMann-Fisher Publishing, Auburn, NH, 2010)

Standard Model Symmetries, And Four and Sixteen Dimension Complex Relativity; The Origin Of Higgs Mass Terms (Blaha Reasearch, Auburn, NH, 2012)

The Bridge to Dark Matter; A New Sister Universe; Dark Energy; Inflatons; Quantum Big Bang; Superluminal Physics; An Extended Standard Model Based on Geometry (Blaha Reasearch, Auburn, NH, 2013)

Available on bn.com, Amazon.com, Amazon.co.uk and other international web sites as well as at better bookstores (through Ingram Distributors).

Preface

In 1961, at the Manhattan YMHA, New York City, I heard Mr. T. S. Elliot say, "Our Beginnings never know our Ends." In physics we may say, "We know our Beginnings from our Present." But finding the Beginnings is a difficult task with many possible branches into the past. One purpose of this book is to find the path to the Beginning within the framework of the multiverse, from the Present state of our universe knowing that there is a strong likelihood of a sister universe. The existence of a sister universe is suggested by the need to generate a fundamental mass (and the inertia of bodies in our universe) for Higgs bosons, a "clock" universe with which to measure time in its deepest sense, and a quantum mechanical Observer which every quantum physical entity requires for measurements. Each universe in the two universe model provides mass, inertia, time, and an Observer for the other universe – just as two siblings often aid each other within a human family.

In the past thirty years or so two main trends of development have been followed by Elementary Particle Theorists to extend our understanding of the basis of Particle Physics. The more popular, top down, approach is based on the assumption of a SuperSymmetry framework and then an attempt to derive the Standard Model and other new features particular to Supersymmetry. The other, bottom up, approach is to explore possible bases for the Standard Model in geometry or other formalisms to uncover the reasons for its seemingly unusual form. In addition many studies have been made of possible extensions of The Standard Model that address CP violation and other issues not currently well understood, or that expand The Standard Model symmetry group to larger, often more exotic symmetries.

In the past twelve years we have pursued a bottom up approach starting from the known features of The Standard Model and sought to find a *direct* underlying basis in the geometry of space-time. This pursuit has led us to propose increasingly deeper and more detailed derivations/constructions of a New Standard Model backed by experiment *and* a fundamental theory.

In our recent studies we have come to realize that the formalism of Logic, particularly Asynchronous Logic, is the only absolutely necessary part of any fundamental physical theory. No theory can exist without Logic to connect the branches of the theory and establish its consequences. Other fundamental approaches require *ad hoc* assumptions that simply raise questions as to their origin and of a possible yet deeper level.

Asynchronous Logic is an appropriate starting point because it addresses the fundamental aspect of particle physics – the interactions, and chains of interactions, between particles. Interactions have a spatial and temporal aspect. In the case of electrical circuits the spatial and temporal evolution of an electric circuit is completely linked due to the known propagation speeds of electric currents in wires. In the case of particle interactions the parts of the interaction process have different speeds of propagation. Thus we need two embodiments of Asynchronous Logic for particle processes – one to coordinate the coordinate motions and one to coordinate the temporal progress of the parts of particle processes.

Asynchronous Logic is a 4-valued logic that can be formalized with 4-vectors and 4 by 4 matrices as we showed in previous books. This immediately implies a 4-dimensional space for particle physics that we assume is complex in anticipation of the need for a complex space-time for tachyon particles – an important part of our New Standard Model. In addition we need another complex 4-dimensional space for the temporal aspect of particle interactions. These paired universes, located in a multiverse and united in a wormhole configuration, provide the basic requirements for the New Standard Model that has exactly the experimentally determined form of the Standard Model: the symmetries, the spectrum of fermions and gauge bosons, the origin of the Higgs mass term, and the origin of time, inertia, and the observer for the multiverse.

We shall see that the geometry of complex 4-dimensional space-time suggest an additional SU(2)⊗U(1) Standard Model symmetry that conveniently can be identified as the source of Dark Matter. In addition a quantization of the coordinates eliminates infinities in its quantum field theory embodiment and leads to a finite Big Bang.

Many parts of our theory are similar to aspects of the multiverse that were derived based on General Relativity and particularly the Wheeler-DeWitt equation. Much of the book is devoted to the extension of the multiverse formalism to complex Euclidean space. It has critical implications for the origin of Higgs particle mass terms and inertia. We also resolve the multiverse Observer problem and the problem of the "Clock" of the multiverse using the sister of our universe.

The Big Bang of our universe with which our previous book ended is now seen as a creation event in the multiverse. A "gauge" field is introduced in the multiverse that can generate paired universes as a vacuum fluctuation. The Wheeler-DeWitt equation is generalized to complex metrics, thus providing a wider framework for the study of the multiverse with a view towards eliminating its problematic points. The generalized Wheeler-DeWitt equation provides an origin for the Cosmological Constant, an origin for the spatial asymmetry of the Universe, and an understanding of the origin for the newly found Web of Galaxies (that links all the groups of galaxies) in our universe.

Some recent books of the author of relevance for the present work are:

The Bridge to Dark Matter; A New Sister Universe; Dark Energy; Inflatons; Quantum Big Bang; Superluminal Physics; An Extended Standard Model Based on Geometry

From Asynchronous Logic to The Standard Model to Superflight to the Stars; Volume 2: Superluminal CP and CPT, U(4) Complex General Relativity and The Standard Model, Complex Vierbein General Relativity, Kinetic Theory, Thermodynamics;

From Asynchronous Logic to The Standard Model to Superflight to the Stars;

Standard Model Symmetries, And Four and Sixteen Dimension Complex Relativity;

The Algebra of Thought & Reality: The Mathematical Basis for Plato's Theory of Ideas, and Reality Extended to Include A Priori Observers and Space-Time;

The Equivalence of Elementary Particle Theories and Computer Languages: Quantum Computers, Turing Machines, Standard Model, Superstring Theory, and a Proof that Gödel's Theorem Implies Nature Must Be Quantum

Quantum Theory of the Third Kind: A New Type of Divergence-free Quantum Field Theory Supporting a Unified Standard Model of Elementary Particles and Quantum Gravity based on a New Method in the Calculus of Variations

Quantum Big Bang Cosmology: Complex Space-time General Relativity, Quantum Coordinates, Dodecahedral Universe, Inflation, and New Spin 0, ½, 1 & 2 Tachyons & Imagyons™ (Pingree-Hill Publishing, Auburn, NH, 2004)

Our later books change, and correct, the earlier books as our understanding of these topics increased. Earlier books often contain theoretical proposals (and some typos) that these more recent books revise based on further thought and study. Some of the earlier books do contain correct details not present in the above mentioned books. The development of the correct form of tachyon quantum field theory in particular, which had not significantly progressed since the work of Sudarshan and Feinberg, was a journey marred by some missteps over the past twelve years. Developing a valid free tachyon quantum field theory is non-trivial as early attempts demonstrate.

CONTENTS

FIGURES and TABLES

1. Summary of the Derivation of our New Standard Model from Space-Time Geometry

Our goal in this chapter is to show the general considerations leading to a New Standard Model – omitting details presented in earlier books by the author. We show that this New Standard Model (which includes the known form of the conventional Standard Model) follows from space-time geometry. In showing this we will extend the conventional Standard Model to include another SU(2)⊗U(1) symmetry with associated gauge fields and fermions that we propose to identify as Dark Matter. We also extend the Standard Model in its internals by identifying neutrinos as tachyons, and quarks as particles having complex 3-momenta. We specify up-type quarks as normal fermions, and down-type quarks as tachyonic fermions.[1]

1.1 Basic Features of Particles and Their Interactions

We know that elementary particles exist and interact with each other in specific ways. The basic features of particles are their existence as discrete entities, their location in space and time, and their interactions (mechanisms of change). In the following sections we will examine the details of these features as we understand it theoretically, and as we see much of it experimentally, in a New Standard Model.[2]

---

[1] This characterization of up and down quarks as "normal" and tachyonic respectively remains to be confirmed experimentally as does their complex 3-momenta. However it appears to be the natural choice.

[2] Recent findings that an electron appears perfectly spherical, and without a detectable dipole moment, to an incredible degree of accuracy – suggesting very strongly that the known particles of the Standard Model are truly elementary.. See the ACME collaboration results. (The ACME group is led by David DeMille (Yale), and John Doyle and Gerald Gabrielse (Harvard). They found no signs of an electron electric dipole moment. The electron seems spherical to within 0.00000000000000000000000000001 centimeter. See J. Baron et al, arXiv:1310.7534v2 (2013).

1.2 Inherent Nature of Elementary Particles

In most particle physics theories the fundamental nature of the truly elementary particles is usually stated in mathematical terms that furnish an answer of sorts but leave open the question of what the mathematics represents. To say that they are strings is an answer but that is still a mathematical characterization – a *mathematical* string – leaving open the question of the composition of the string and whether they have more fundamental constituents. After all, strings have parts.

To say that particles are really "wave-particles" with certain properties also leaves open the question of their inherent composition.

There appears to be only one view of the nature of elementary particles that provides an "ultimate" solution to the question of the composition of elementary particles: elementary particles are logic units that have attached properties (position, momentum, mass, and quantum numbers) that originate directly or indirectly from space-time properties. In this view particle spin is a logic value although it has the form and features of a physical property.[3]

The concept of spin as a logic value,[4] as the core feature of an elementary particle, enables us to view an elementary particle in a way that precludes the possibility of a yet lower level of fundamental particles thus ending a chain of successive lower levels of Reality. Again we note Logic is the only absolute requirement of any non-trivial physical theory. All physics theories derive their results via logic – without which nothing.

Thus we developed our fundamental theory firmly based on Logic – Asynchronous Logic in particular – which *gives a reason for 4-dimensional space-time, and presents a preliminary justification for two 4-dimensional universes: our universe and a second sister universe that is the clock,*[5] *the source of mass*

---

[3] Spin zero particles have only one truth value: true. Spin ½ particles have four truth values. And so on for higher spin particles. Due to a similar mathematical formulation for angular momentum spin can be combined with angular momentum as if it were a form of inherent angular momentum. Clearly the direct interpretation of spin as "spinning" is not correct since it would, for example, have electrons spinning faster than the speed of light. (Note: Spin 0 particles have the value true; their anti-particles have the value false.)

[4] Our concept clearly has a precursor in the monads of Leibniz. See Rescher (1967) for a discussion of monads.

[5] DeWitt, B. S., Phys. Rev. **160**, 1113 (1967) seems to consider the possibility of a universe as a clock on p. 1137.

and inertia of our universe, and the multiverse "observer" of our quantum universe.[6]

We will be led to suggest that these universes are connected by a wormhole. These developments will provide a basis for physical reality. The choice of dimensions and universes will then not be a matter of random speculation unconnected with nature. We will also use a multiverse formulation (extended beyond the standard multiverse to a complex metrics multiverse) which appears to provides an origin for the Cosmological Constant, an origin for the spatial asymmetry of the Universe, and an understanding of the origin for the newly found Web of Galaxies (that links all the groups of galaxies) in our universe. It also provides a framework for a theory of the Big Bang origin of our universe. This precursor formulation leads to the Quantum Big Bang theory of Blaha (2004) as presented in Blaha (2013).

1.3 Implications of Characterizing a Particle as a Logic State with Space-Time Added

Logic statements often require the specification of space-time coordinates within the statement.[7] Elementary particles are located at space-time points (although their exact locations must be determined by observations.) Particles experience a variety of interactions in space-time via the exchange of gauge bosons and Higgs particles. High order perturbation theory calculations show that interactions between particles can take place with large temporal and spatial separations. Spatially and temporally separated Interaction processes are synchronized by the dynamics of quantum field theory.[8]

---

[6] Solving an issue discussed in DeWitt (p. 1131) who viewed the lack of a multiverse observer as a deviation from the Copenhagen interpretation of Quantum Theory. We view the sister universe as the theoretical observer of our universe thus placing the quantum multiverse back within the Copenhagen interpretation. We discuss this concept in more detail later.

[7] For example, "The car is in the garage."

[8] An obvious, and thus little thought of, feature of elementary particle phenomena is the coordination of the parts of a physical process in time and space. Complex Feynman diagrams embody the coordination of the spatially separated parts of interacting particles over a period of time. Quantum entanglement embodies the coordination of the parts of a physical phenomenon separated by small and large distances. These examples, which could be multiplied indefinitely, lead to Physical Principles of Spatial and Time Asynchronicity.

The spin states of fermions can be viewed as logic states in 4-valued and 3-valued matrix representations of logic.[9] Changes of spin in interactions correspond to changes in logic values. Thus we view particles as logic units combined with features derivative from space-time geometry.

In Blaha (2013), and earlier work, we showed that Asynchronous Logic[10] was necessary for the coordination of phenomena that had spatial and/or temporal separations. We defined two principles of Asynchronicity:

Asynchronicity Principle 1: Nature requires spatial coordination of physical phenomena through the use of Asynchronous Logic.
Asynchronicity Principle 2: Nature requires time coordination of physical phenomena through the use of Asynchronous Logic.

Asynchronous Logic is a 4-valued logic.[11]

In the case of physical phenomena temporal and spatial progress are separate coordination issues[12] (although resolved by quantum field theory)

---

[9] Gravitons have spin 2. Their spin states can also be viewed as logic states.

[10] *Asynchronous Logic* is the logic used in the design of asynchronous circuits. This logic embodies asynchronicity, and so circuits built using it do not use a clock to control the execution speed of the various parts of an asynchronous circuit. Consequently logic elements do not necessarily have a distinct true or false state at any given point in time. This logic supports "stop and go" states within an executing asynchronous circuit. Time and space are coordinated in electrical circuits due to a constant speed of electrons through circuits. Otherwise VLSI design would need separate time and space synchronization. In earlier 2-valued logic circuit design approaches the spatial synchronization would be governed by a separate clock. This situation – of clocks vs. multi-valued logic – invites comparison with Wheeler-DeWitt equation dynamics where the spatial evolution of the metrics is governed by a separate external clock, or by introducing a separate universe that serves the role of a timing mechanism. See p. 1137 of DeWitt, B. S., Phys. Rev. **160**, 1113 (1967) where the possibility of using an entire universe as a clock is considered, and also p. 1059 of Unruh, W. G., Phys. Rev. **D40**, 1053 (1989). The author was unaware of these multiverse considerations when he formulated his Asynchronicity Principles in Blaha (2012) and Blaha (2013). *The similarity of concepts is a happy coincidence that we will consider in detail in later chapters together with other similarities that lead us to see our work as providing the first evidence of the reality of the multiverse.*

[11] In Fant's asynchronous 4-valued logic the four possible truth values are: True – status is true and all data is current, False – status is false and all data is current, Intermediate – status is indefinite with some data current, and NULL – status is indefinite with no data present – results in a suspension of processing of the circuit part in a NULL state until current data becomes present. A 2-valued Logic formulation combined with a constraint condition is also possible just like the Dirac equation can be replaced with a Klein-Gordon equation and a supplementary condition. See Fant (2005).

[12] Time and spatial asynchronicity are different in the case of particles because massive particle propagators do not have a fixed relation between time and distance in their propagation. In VLSI circuits the time

because the spatial evolution of phenomena is different in principle from the time evolution.[13]

Due to separate coordination issues we determined a need for two 4-valued logic spaces.[14] One of these logic spaces will lead to the space-time of our universe. The other logic space will lead to the space-time of another universe that we have called the *sister universe*.

We suggested previously that the above Principles of Asynchronicity are implicitly embodied in our New Standard Model. We showed that these principles have major consequences in two areas: 1) they lead to Dirac-like equations for the fundamental fermions – leptons and quarks – in our universe; and 2) they also imply four generations of fermions due to our sister 4-dimensional universe.

The matrix form of a 4-valued Asynchronous Logic (which we tentatively assume is spatial[15] Asynchronous Logic) leads us to identify Logic 4-vectors with Dirac spinors implying a 4-dimensional space-time[16] for our universe. Thus we are led to:

---

interval between nodes is fixed by the distance between the nodes since the "wire" connecting them determines the speed of transmission. Thus we need two universes, our universe and a sister universe, for particle asynchronicity but only one "universe" of discourse for VLSI circuits.

[13] Again we note: in the case of electrical circuits such as VLSI chip circuits, time and space behavior are related by the speed of propagation of electrons in electric currents. Thus only one synchronization procedure is needed. In particle interactions the "speed" of propagation of effects within a process is indeterminate and thus spatial and time synchronization are separate issues in general. Quantum theories of the multiverse particularly reflect this.

[14] A logic space is a matrix representation of a Logic. For example a 4-valued logic has a 4 by 4 matrix representation.

[15] Formerly, it appeared that this choice was arbitrary as far as the state of our understanding of the Standard Model was concerned. However if one considers the Wheeler-DeWitt equation (DeWitt, B. S., Phys. Rev. **160**, 1113 (1987)) and the goal of determining time from 3-dimensional geometry then it would be more natural to consider spatial asynchronicity as the basis of our 4-dimensional universe. (This view is the opposite of the conjecture in Blaha (2013) which did not take general relativistic considerations into account. See Baierlein, R. F., Sharp, D. H., and Wheeler, J. A., Phys. Rev. **126**, 1864 (1962).) We consider the emergence of time from 3-geometries later in more detail.

[16] Weinberg (1995) p. 216 exhibits an equation that relates the number of components of a spinor to the dimension of its space-time. This equation shows that a 4 component logic vector when viewed as a spinor (under a mapping of logic values to spinor components) implies a 4-dimensional real or complex space-time. We will require a *complex space-time* in order to support superluminal transformations and tachyon particles. Tachyon particles are necessary to derive the form of The New Standard Model.

1. Fermion (matter) particles have four fundamental states corresponding to 4-valued Logic states
2. They are spin ½ particles in 4 dimensions. A four-valued logic state vector is a spinor in 4 dimensions (with which coordinates are associated later.)
3. They have 4×4 dynamical equations upon the introduction of coordinates that make spinors local.

We assume temporal Asynchronous Logic[17] – also 4-valued – leads to a sister universe that endows our universe with four generations of fermions. This universe is similarly a 4-dimensional universe which, based on a sense of symmetry, we take to have the same physical laws and space-time structure as our universe. The evolution of the sister universe may be different, and its distribution of matter is most likely different, but there is no apparent reason for other differences from our universe.

1.4 Complex 4-Dimensional Space-Times

Having established two 4-dimensional universes we now contend that the space-times must be complex. As we have shown in earlier work there is good reason to believe that certain types of elementary particles, namely neutrinos and down-type quarks are tachyons – faster than light particles. If this is so, then Lorentz transformations between a lab frame's coordinates and a tachyon's rest frame coordinates must be complex-valued in general. Thus accommodating tachyons in our Standard Model, as we must do to achieve the known form of The Standard Model, necessitates the complex coordinates and transformations of the complex Lorentz group[18] for our universe and for our sister universe.

It then becomes necessary to define invariant intervals in each space-time separately with the form:

$$ds^2 = g_{\mu\nu}dx^\mu dx^\nu \qquad (1.1)$$

---

[17] Formerly in Blaha (2013) we assumed the sister universe followed from spatial Asynchronous Logic. General Relativistic considerations (related to the Wheeler-DeWitt equation) have changed our thinking to this new view.

[18] This feature accords well with axiomatic quantum field theory. As Streater (2000) shows axiomatic quantum field theory proofs require a *complex* Lorentz group.

where the metric and coordinates are complex in general. Eq. 1.1 implies that

$$g_{\mu\nu} = g_{\nu\mu} \tag{1.2}$$

1.5 The Reality Group and Measurements of Space-Time Coordinates

The reader is well aware that measurements of time with clocks, and measurements of spatial coordinates, always yield real numeric values. Thus in a valid coordinate system there must be a mechanism to extract a real value for each point if the point's coordinates are complex. To that end we introduced[19] a local group of U(4) transformations, which we call the *Reality group*, whose set of transformations transform the coordinates of any point in a complex 4-dimensional space-time to four real values. Thus eq. 1.1, for example, takes a more conventional real-valued form after the application of a local U(4) transformation to the displaced coordinates:

$$ds^2 = g'_{\mu\nu}dx'^{\mu}dx'^{\nu} \tag{1.3}$$

where the metric $g'_{\mu\nu}$ and the displacements dx'^{μ} are real.

Remarkably the subgroups of U(4) are the same as the known groups of the Standard Model: SU(3) and SU(2)⊗U(1) with one addition: another SU(2)⊗U(1) subgroup[20] that is not yet experimentally found but which we believe is the group of Dark Matter – yielding an additional four heavy gauge vector bosons and twelve[21] fermions that we identify with Dark Matter and their primary interactions.

The U(4) Reality group has 16 generators. In four dimensions it is not possible to have separate commuting U(4) subgroups (SU(3), SU(2)⊗U(1), and another SU(2)⊗U(1)). The idea of using these subgroups *directly* as the origin of the Standard Model symmetries, while attractive, does not work because the

---

[19] Blaha (2012) and Blaha (2013).

[20] Blaha (2013) describes this additional symmetry's features in detail: lagrangian terms, particle spectrum and the recent CERN experiments generating excess electron-positron pairs attributed to Dark Matter.

[21] Assuming three generations, and realizing Dark Matter does not have the strong interactions, so Dark quarks are color singlets. See Blaha (2013).

group symmetry, R = SU(3)⊗SU(2)⊗U(1)⊗SU(2)⊗U(1), of the extended Standard Model has commuting factors.[22]

The simplest representation of the factors of R in which the factors are commuting groups, and which can have a fully reducible[23] representation, is the 16 dimensional representation of U(16). Its 16-dimensional complex space supports a fully reducible R group representation. A more important reason for focusing on 16-dimensional complex space is presented in the next section.

1.6 Embedding Curved Space-time in a 16-Dimensional Space

Although The Standard Model is defined in a flat 4-dimensional space-time it appears that the universe is actually curved and in fact closed. We can embed our universe within a larger universe. If our universe were a 4-dimensional real universe, then its metric $g_{\mu\nu}$ has 10 independent components that are determined by the Einstein dynamic equations. Thus we could embed a real universe as a surface in a 10-dimensional flat space as Eddington (1952) pointed out.

However a 4-dimensional *complex* universe has a metric $g_{\mu\nu}$ with 16 independent real-valued components in view of eq. 1.2. If one embeds our complex 4-dimensional curved space-time within a Euclidean flat space, then the flat space must be at least a 16-dimensional real space specified by sixteen equations:

$$z_i = f_i(x) \tag{1.4}$$

where x is a complex 4-vector in our universe[24] (as in chapters 2 and 3 of Blaha (2012b)) and z_i for i = 1, ..., 16 are the coordinates of a flat space point. The functions f_i map our universe into a 16 dimensional real, flat space as a complex 4-dimensional surface (an 8-dimensional real-valued surface.)

---

[22] Some physicists have proposed an ultimate unification of all symmetries at some high energy based on an extrapolation of low order perturbation calculations. However there is no evidence that these commuting group factors will then become non-commuting. The origin of the various groups that constitute the Reality group in 4 dimensions shows that each performs a different role – see appendix 18-A and chapter 19 of Blaha (2011c).

[23] See Blaha (2013) p. 32 or earlier books.

[24] Our universe is then a complex 4-surface within a 16-dimensional complex flat space.

We now point out that the 16-dimensional real space should be replaced with a 16-dimensional *complex* flat space that we have called the Flatverse in Blaha (2012b) with invariant interval

$$ds^2 = g_{ij}dz^i dz^j \tag{1.5}$$

with the Flatverse metric satisfying

$$g_{ij} = g_{ji} \tag{1.6}$$

The metric tensor $g_{\mu\nu}$ of the universe can be defined in terms of Flatverse coordinates by

$$g_{\mu\nu} = \partial f_j/\partial x^\mu \, \partial f_j/\partial x^\nu \tag{1.7}$$

with an implied sum over the subscript j.[25]

The complex 16-dimensional Flatverse can also have a Reality group defined for it. After all, if there were multiple universes within the Flatverse, then distance measurements between universes would necessarily be real-valued. We define the Flatverse Reality group as G_{z16} = R = SU(3)⊗SU(2)⊗U(1)⊗SU(2)⊗U(1). One can easily show that the 16 dimension Reality group G_{z16} that corresponds to the 4-dimensional Reality group, denoted G_{x4}, must has 16 generators. (The other 16 generator possibilities are SU(3)⊗SU(3) and U(4). U(4) is unacceptable because its generators do not have appropriate commutation relations – they would mix the strong and ElectroWeak interactions. SU(3)⊗SU(3) has 16 generators—but the commutation relations of this group do not conform to the physical roles of the generators of the Reality group for 4 dimensional transformations. See appendix 18-A and chapter 19 of Blaha (2011c).) Thus G_{z16} is the only reasonable choice for a 16 dimension Reality group.

The proof that G_{z16} is the correct choice for the 16 dimension Reality group follows from eq. 1.4. Suppose we perform a Reality group transformation G_{x4b} on a 4-dimensional vector in our universe. Then there must be a

---

[25] This form represents a return to the complex form of invariant intervals of Blaha (2004). An alternate form of complex invariant intervals and metrics presented in Blaha (2012b) does not appear to be of interest after some study by this author.

corresponding Reality group transformation G_{16a} of the flatverse coordinates. So we can write

$$y = G_{z16a}z = f(G_{x4b}x) \quad (1.8)$$

where y is a real 16-vector, G_{z16a} is an element of G_{z16}, G_{x4b} is an element of G_{x4}, z is a 16-dimensional vector and f is a vector composed of the 16 f_i functions. The 16 dimension group G_{z16} has 16 generators that we will denote V_i. The 4 dimension Reality group G_{x4} = U(4) has 16 generators that we will denote U_i. If we make an infinitesimal transformation G_{x4b}

$$G_{x4b} = I + \beta_i U_i \quad (1.9)$$

then there must be a corresponding infinitesimal transformation G_{z16a}

$$G_{z16a} = I + \alpha_i V_i \quad (1.10)$$

where α_i and β_i are constants[26] for i = 1, 2, ..., 16. Substituting in eq. 1.8 and expanding to first order we find

$$\alpha_i V_{ijk} f_k = \beta_i U_i{}^{\mu}{}_{\nu}\, x^{\nu}\, \partial f_j/\partial x^{\mu} \quad (1.11)$$

for j = 1, 2, ..., 16 with summations over i, k, μ, and ν. Eq. 1.11 is 16 equations that determine the β_i parameters in terms of the α_i parameters. Thus a G_{x4} transformation uniquely determines a G_{z16} transformation.

In the case of a flat 4 dimensional space-time where we can limit the Flatverse to 4 dimensions also ($z_i = x_i$ for i = 1, 2, ...,4), then the 16 dimension Reality group must be U(4) (not G_{z16}) and eq. 1/11 simplifies to[27]

$$\alpha_i V_{ijk} = \beta_i U_{ijk} \quad (1.12)$$

---

[26] They could be functions of z and x respectively in the case of a curved 4-dimensional space-time or for accelerating reference frame transformations.

[27] We will not distinguish between raised and lowered indices for the sake of simplicity as they do not have physical import in these considerations.

implying $\alpha_i = \beta_i$ and $V_{ijk} = U_{ijk}$. Thus the Flatverse's and our universe's Reality transformations coincide for a flat universe. In the case of a curved universe eq. 1.12 requires a more complex calculation. In any case, the equality of the number of Reality group generators in our universe and the Flatverse is crucial. Otherwise a solution is either ambiguous or non-existent in general.

An additional point of importance is that the Flatverse must also be Euclidean in order to support quantum gravity[28] as we discuss in detail later.

1.7 Covariant Derivatives and Connections in the Flatverse

The introduction of the Reality group in the Flatverse leads to the need for covariant derivatives for quantities that are subject to Reality group transformations. This question has been addressed in Blaha (2012a). In this section we will introduce covariant derivatives necessitated by the Reality group.

Consider a 16 component vector function of the coordinates of the Flatverse **F**(z). If we apply a Reality group transformation G_{z16a} to it (to make all its components real-valued) then the partial derivative of $G_{z16a}\mathbf{F}$ changes in a non-covariant way.

$$\partial(G_{z16a}\mathbf{F})_j/\partial z^i \neq G_{z16ajk}\partial(\mathbf{F})_k/\partial z^i \tag{1.13}$$

In order to have a covariant derivative expression we must take the 16 generators of G_{z16}, which we denote V_i for i = 1, 2, ... , 16, and define the covariant derivative

$$D_k = \partial/\partial z^k - iV_iZ_{ki} \tag{1.14}$$

where the quantities Z_{ki} are connections (using the terminology of General Relativity), and where there is an implicit sum over i. Using the covariant derivative we find

$$D_k(G_{z16a}\mathbf{F}) = G_{z16a}\partial\mathbf{F}/\partial z^k \tag{1.15}$$

---

[28] Hawking, S. W., Phys. Rev. D **37** 904 (1988).

A similar need for covariant derivatives appears in 4-dimensional Lorentzian space-time. In this subspace the Reality connections lead to gauge fields that embody the interactions of the Standard Model.

1.8 Covariant Derivatives in 4-Dimensional Lorentzian Universes such as Our Universe

In our complex 4-dimensional Lorentzian universe

$$ds^2 = {}_4g_{\mu\nu}dx^\mu dx^\nu \quad (1.16)$$

where ${}_4g_{\mu\nu}$ is the complex conventional Lorentz metric with the metric satisfying

$${}_4g_{\mu\nu} = {}_4g_{\nu\mu} \quad (1.17)$$

We can use the 4×4 U(4) fundamental matrix representation of the Reality group to transform complex coordinates to real-valued coordinates. However the subgroups of U(4) do not commute making their direct use as particle symmetry groups impossible because the Standard Model groups: the ElectroWeak interactions group, the strong interaction group, and a new SU(2)⊗U(1) group, that we have suggested is the Dark Matter interactions group, all must commute.

Again we define a covariant derivative similar to that of eq. 1.14. Now the 4-dimensional space-time Reality group of the New Standard Model is R = SU(3)⊗SU(2)⊗U(1)⊗SU(2)⊗U(1), a direct product group. It performs a role similar to the 16-dimensional case (eq. 1.8):

$$w = R_1x \quad (1.18)$$

where w is a real-valued 4-vector and R_1 is an R transformation. The *specific* form of the Reality group is derived in chapters 18, 18A, and 19 of Blaha (2011c) and chapter 1 of Blaha (2013) from a consideration of complex Lorentz group transformations. As expected the 4-dimensional Reality group has 16 generators, being a local version of R and so it can transform any complex 4-vector into a real-valued vector. Since it is independent of the complex Lorentz group the "noGo" theorems of Coleman and others do not apply within this framework –

there is no unification of the Reality group and the Lorentz group such as appeared, for example, in the case of SU(6).

Due to the Reality group, the covariant derivative for the New Standard Model must have the form

$$D_\mu = \partial/\partial x^\mu - iV_{8i} A^i_\mu \tag{1.19}$$

where A^i_μ are 16 gauge fields - connections (i = 1, ..., 16), and the V_{8i} are the 16 generators of SU(3)⊗SU(2)⊗U(1)⊗SU(2)⊗U(1).

We note that we are required to use a *complex* 4-dimensional space-time to accommodate tachyons, and their related complex Lorentz transformations. There is also the demonstrated need for a complex 4-dimensional space-time in axiomatic quantum field theory (See Streater (2000).)

1.9 New Standard Model Gauge Fields

The connections listed in eq. 1.17 are the covariant derivatives that appear in our extended Standard Model. Their role is described in detail in the extended Standard Model in Blaha (2011c) and the extension of The Standard Model to include SU(2)⊗U(1) Dark Matter[29] is described in chapter 1 of Blaha (2013).

Reality group transformations U, when applied to wave functions $\psi(x)$,

$$U\psi = D\psi \tag{1.20}$$

have representations, D, that lead to the fermion spectrum and the familiar interactions of The Standard Model. The New Standard Model is derived in detail in Blaha (2011c) and Blaha (2013).

An important requirement to achieve the form of the Standard Model is the appearance of tachyons: tachyon neutrinos and tachyon down-type quarks. We show tachyons have an acceptable quantum field theory in Blaha (2011c) and earlier books. We also show faster-than-light kinetic theory and thermodynamics are physically reasonable in Blaha (2012a).

---

[29] Which are also among the 16 connections in eq. 1.17.

Blaha (2011c) and (2013) describe the lagrangian terms, the Faddeev-Popov ghosts and other features of these gauge fields. Thus the gauge field sector of the New Standard Model is fully described.

Other aspects of the extended Standard Model: fundamental fermions and Higgs particles are also described in Blaha (2013).[30] We summarize these results below.

1.10 Fermion Spectrum

The known fundamental fermions, quarks and leptons, have a spectrum consisting of four species: neutrinos, charged leptons, up-type quarks and down-type quarks. All quarks appear in triplets due to color SU(3) symmetry. Each species currently has three generations. In our New Standard Model we provisionally add a fourth generation to each species.[31]

Due to its additional SU(2)⊗U(1) symmetry there are four additional species of Dark Matter: Dark neutrinos, Dark non-electric charged leptons, Dark up-type quarks and Dark down-type quarks. Dark quarks are color SU(3) singlets since they do not interact with normal matter via the strong interactions. Their charges are non-electric.

1.10.1 Origin of Four Species of Fermions

The four quark species are due to the complex Lorentz group of our space-time, and *the requirement that a pure Lorentz boost of a particle from its rest frame boosts to a frame in which the particle has a real-valued energy.* Fundamental particles are stable in the absence of interactions and thus must have a real-valued energy.

The most general complex Lorentz boost has the form

$$\Lambda_C(\mathbf{v_c}) \equiv \Lambda_C(\omega, \mathbf{v_c}) = \exp[i\omega\hat{\mathbf{w}}\cdot\mathbf{K}] \qquad (1.21)$$

where K is the boost vector and where the other parameters are defined by

$$\omega = (\omega_r^{\,2} - \omega_i^{\,2} + 2i\omega_r\omega_i\,\hat{\mathbf{u}}_r\cdot\hat{\mathbf{u}}_i)^{1/2} \qquad (1.22)$$

[30] Earlier versions appear in Blaha (2011c), (2012a) and (2012b).

[31] Preliminary experimental evidence for a fourth generation is beginning to appear.

$$\hat{\mathbf{w}} = (\omega_r\hat{\mathbf{u}}_r + i\omega_i\hat{\mathbf{u}}_i)/\omega$$
$$\hat{\mathbf{u}}_r\cdot\hat{\mathbf{u}}_r = 1 = \hat{\mathbf{u}}_i\cdot\hat{\mathbf{u}}_i$$
$$\hat{\mathbf{w}}\cdot\hat{\mathbf{w}} = 1$$

The velocity is

$$\mathbf{v}_c = \hat{\mathbf{w}}\tanh(\omega) \tag{1.23}$$

The four cases that yield a particle with a real-valued energy and the corresponding fermion species are:

1. Charged Leptons (both normal and Dark charged)

$$\hat{\mathbf{u}}_i = 0 \text{ and } |\mathbf{v}_c| < c$$

2. Neutrinos (both normal and Dark) - tachyons[32]

$$\hat{\mathbf{u}}_i = 0 \text{ and } |\mathbf{v}_c| > c$$

3. Up-type Quarks (both normal and Dark)[33]

$$\hat{\mathbf{u}}_i \neq 0 \text{ and } |\mathbf{v}_c| < c$$

4. Down-type Quarks (both normal and Dark) - tachyons

$$\hat{\mathbf{u}}_i \neq 0 \text{ and } |\mathbf{v}_c| > c$$

Thus we have a simple explanation for the existence of exactly four species of normal fermions and of Dark fermions. See Blaha (2011c) for more details.

1.10.2 Differentiation Between Quarks and Leptons

The previous subsection shows the difference between leptons and quarks is in their 3-momenta: leptons have real-valued 3-momentum and quarks

---

[32] The evidence for the existence of tachyons is discussed in chapter 5 of Blaha (2013) as well as earlier books.

[33] Whether up-type quarks are normal and down-type quarks are tachyons or vice versa has not been decided experimentally. So the situation may be reversed in future.

have complex-valued 3-momenta. The extra degrees of freedom in quark 3-momenta lead to their color SU(3) symmetry. This symmetry which is part of the Reality group is dynamically significant although any direct measurement of a quark's momentum will yield a real value. Chapter 19 of Blaha (2011c) discusses the symmetry in detail.[34] Normal quarks are in the fundamental representation $\underline{3}$ of color SU(3). Dark quarks are color singlets (the $\underline{0}$ representation of SU(3)) since they do not experience the strong interaction.

1.10.3 ElectroWeak and Dark ElectroWeak Doublets - Origin of Parity Violation

The SU(2)⊗U(1) ElectroWeak symmetry of normal matter leads to a doublet structure of fermions.[35] This structure can also be derived through transformations (subsequently broken) that rotate tachyons into normal particles and vice versa. These transformations are described in detail in Chapter 18 of Blaha (2011c).

The same type of transformations apply to Dark SU(2)⊗U(1) "ElectroWeak" symmetry and so Dark doublets consist of a Dark charged lepton and Dark neutrino, and of a Dark up-type quark and Dark down-type quark. (See Blaha (2013).)

The form of ElectroWeak doublets consisting of a "normal" fermion and a tachyon leads directly to Parity Violation as pointed out in Blaha (2011c) and our earlier work. Thus the origin of Parity Violation has finally been found.

1.10.4 Origin of Four Generations

The origin of the four generations (only three of which have been found as yet) is due to the existence of a sister universe that is motivated by temporal Asynchronous Logic. The sister universe has the same overall structure as our universe – in particular it is a complex 4-dimensional Lorentzian space-time. We will discuss it in detail in subsequent chapters. The 4-dimensional nature of the sister universe leads to four component spinors that are the direct source of the four generations. Following pp. 282-286 of Blaha (2011c), except that we insert a

---

[34] Again we note that NoGo theorems do not apply in this case because the symmetry is not joined with the Lorentz group.

[35] Normal matter doublets consisting of a charged lepton and neutrino; Dark Matter doublets consisting of a Dark "Charged" lepton and a Dark neutrino.

Higgs field coupling now that Higgs particles have been found, we start with a unified Dirac-like dynamical equation for our universe and its sister universe where D… represents a derivative term together with an interaction term

$$[D_U \otimes I_S + I_U \otimes D_S + g_S \Phi(x, y) M] \psi_R(x, y) = 0 \quad (1.24)$$

where U represents our universe, S represents the sister universe, and where $\Phi(x, y)$ is a Higgs field in both universes as explained in Blaha (2013) and in chapter 2 in more detail. I is the identity matrix. Explicitly, with indices displayed, and the D terms expanded, eq. 1.24 is[36]

$$[\gamma^{\nu}{}_{ab} \mathcal{D}_{\nu}(x) + \gamma^{\nu}{}_{cd} \mathcal{D}'_{\nu}(y) + g_S \Phi(x, y) M_{cd}\, \delta_{ab}] \psi_{Rbd}(x, y) = 0 \quad (1.25)$$

Upon spontaneous breakdown the Higgs field becomes a constant (plus an additional variable Higgs field part not displayed) which we normalize to unity

$$g_S \Phi(x, y) \rightarrow g_S \quad (1.26)$$

Eq. 1.25 with eq. 1.26 inserted is manifestly separable so we can let

$$\psi_{Rbd}(x, y) = \psi_{Rbd}(x) \psi'_{Rd}(y) \quad (1.27)$$

and separate the equations:[37]

$$\gamma'^{\mu}{}_{cd} \mathcal{D}'_{\mu}(y) \psi'_{Rd}(y) = 0 \quad (1.28)$$

$$(\gamma^{\nu}{}_{ab} \mathcal{D}_{\nu}(x) \delta_{cd} + g_S M_{cd} \delta_{ab}) \psi_{Rbd}(x) = 0 \quad (1.29)$$

or, in matrix notation, eq. 1.29 – the equation for the fermion generations – in our universe becomes

$$(\gamma^{\nu} \mathcal{D}_{\nu}(x) + g_S M) \psi_R(x) = 0 \quad (1.30)$$

---

[36] The ' such as that on $\mathcal{D}'$ indicates the sister universe.

[37] This separation enables the fermions of the sister universe to have a zero mass (or bare mass) as opposed to a mass matrix that is not a multiple of the identity matrix in its Dirac-like equation.

where $\psi_R(x)$ is four generations of a fermion species' right-handed spinors. Thus the fermion generations emerge from geometry – the impact of the sister universe on our universe. We return to this discussion in chapter 2.

1.10.5 Origin of the Higgs Lagrangian Mass Term

In Blaha (2013) we pointed out that the Higgs sector of the Standard Model lagrangian had a Higgs mass term with the "wrong" sign in the sense that it made the Higgs particle a tachyon. The tachyonic nature was submerged due to the quartic interaction term which together with the mass term produced a spontaneous breakdown. We considered a simplified example of the generation of Higgs particle mass terms for the case of complex scalar particle fields.[38] The realistic case of multiplets of Higgs particles is a direct generalization. The lagrangian density has the form

$$\mathscr{L}(\phi(x)) = \partial^\mu\phi\partial_\mu\phi^* - V(\phi, \phi^*) \tag{1.31}$$

where

$$V(\phi, \phi^*) = m^2\phi\phi^* + \beta(\phi\phi^*)^2 \tag{1.32}$$

If $\beta > 0$ and $m^2 < 0$ then spontaneous symmetry breaking occurs and the ϕ field acquires a constant part ϕ_0

$$\phi(x) = \phi_0 + \delta\phi(x) \tag{1.33}$$

with

$$\phi_0 = -m/\surd\beta \tag{1.34}$$

We then went on to show that the mass term in eq. 1.32 is either an ad hoc constant or might reflect a deeper theoretical construct. In particular, we showed that if the Higgs particle was partly in our universe and partly in a sister universe then the mass term might arise as a separation constant when the parts of the dynamic equation are separated for each universe. The mechanism is

[38] Chapter 1 of Blaha (2012b) contains this discussion and example, which we place here for the sake of completeness.

analogous to the appearance of a mass in the Schwarzschild solution in General Relativity.

Since the Higgs mass term sets the scale for all masses in The Standard Model, and in General Relativity, we can attribute the origin of not only the mass scale but also of inertia to the existence of a sister universe. We will discuss this in more detail in chapter 2 together with other important roles of the sister universe, such as the fermion generations described above, that lead us to conclude there are at least two universes, and thus physical reality is truly a multiverse. Fermion generations and Higgs masses are the first partial evidence for a multiverse. Until now the multiverse was mere speculation.

1.11 Dark Matter in the New Standard Model

As discussed earlier in this section the source of Dark Matter is a Dark $SU(2)\otimes U(1)$ "ElectroWeak" symmetry present in the Reality group. It leads to Dark doublets consist of a Dark charged lepton and Dark neutrino, and of a Dark up-type quark and Dark down-type quark. It also leads to four Dark "ElectroWeak" gauge bosons. (See Blaha (2013).)

1.12 Elimination of Infinities in Standard Model Calculations

Blaha (2005a) and earlier books showed a new method to eliminate the infinities that appear in Feynman diagram calculations in perturbation. The method q-number coordinates

$$X^\mu(z) = z^\mu + i\, Y^\mu(z)/M_c^{\,2} \tag{1.35}$$

(where M_c is a very large mass scale of perhaps the order of the Planck mass, and where $Y^\mu(z)$ is a free QED-like field). We called this type of quantum field theory a Two-Tier quantum field theory.

In chapter 4 of Blaha (2012b) we showed that the Two-Tier formulation of a quantum field theory could be viewed as an embedding our universe in the Flatverse for the special case of a flat universe which would allow the chosen Flatverse to be 4 dimensional. The generalization to a curved space-time embedded in a 16 dimension Flatverse is direct.

We also showed in chapter 5 of Blaha (2005a) that a particle field in z^μ coordinates can be "dressed" to be a particle field in $X^\mu(z)$ coordinates.

Thus one can view the Flatverse edge of our universe as the "source" of bare particles which are dressed to become cloaked particles in our universe with q-number coordinates. The cloaking of particles surrounds them with a cloud of $Y^{\mu}(z)$ vector bosons that effectively smears the particles in such a way as to suppress infinities in perturbation calculations.[39] Basically every quantum field propagator has a Gaussian factor suppressing high momentum within it. This factor eliminates infinities in perturbation theory without the use of any renormalization technique. Thus the New Standard Model is free of any infinities.

1.13 Quantum Big Bang

The current state of our knowledge of the evolution of the universe has now been extended back in time to about 350,000 years after the Big Bang through recent astrophysical research. While this progress is encouraging we still face a major issue: the origin of Dark Energy, and the events of those critical years before the 350,000 year point that we have now apparently reached experimentally. Those early years and the Big Bang itself remain mysteries. This situation is especially critical since the early years of the universe apparently contain an uncertain beginning and explosive growth.

In Blaha (2004) and (2013) we presented a quantum Big Bang theory that explains the unknown period in the neighborhood of t = 0 where quantum effects played a major role. It also explains the inflationary growth of the universe, which is usually attributed to an unknown "particle" called an inflaton. We found that the "inflaton" appears to be the energy of the q-number part of quantum coordinates – the quantum field denoted $Y^{\mu}(x)$ in eq. 1.35.

The inflaton makes the Big Bang finite—no singularity; and frees The New Standard Model and Quantum Gravity[40] of infinities. *$Y^{\mu}(x)$ has a remarkable triple role in our view – to eliminate the Big Bang singularity, to generate the explosive growth of the universe, and to remove infinities from The Standard Model and Quantum Gravity.*

Since the $Y^{\mu}(x)$ quantum gauge field is a free field (neglecting gravity) the initial state of the universe can be permeated with a large Black Body spectrum

---

[39] See Blaha (2005a).

[40] See Blaha (2011c) and Blaha (2005a) for the removal of infinities in Quantum Gravity.

of quanta of this field as well as quanta of other particles. The total energy of the free $Y^{\mu}(x)$ field within the universe is then the energy that we call Dark Energy—energy that can only influence the universe through its gravitational effects.

1.14 Axioms for the New Standard Model Based on Space-Time Geometry

In Blaha (2011c) we proposed a set of postulates for the derivation of an extended Standard Model. This set of postulates was a bit lengthy and the postulates overlapped one another.

In this section we propose a set of axioms that is shorter and deeper than the postulate set of earlier work. We anticipate that this set of axioms will prove to be incomplete. But it does have the virtue of being more fundamental.

AXIOMS

1. The dimensionality of physical space-time is set by spatial Asynchronous Logic to enable synchronization of physical processes.
2. The metric of this space-time $g_{\mu\nu}$ is the Lorentz metric with three spatial coordinates and one time coordinate.
3. Space-time coordinates are complex-valued.
4. There is a local unitary group, the Reality group, which transforms complex-valued space-time coordinates to the real-valued coordinates that are measured in experiments.
5. Matter consists of particles that embody Logic values that we call spin states.
6. Particle dynamics are specified by quantum field theory. The dynamic equations derived from quantum field theory are covariant under Reality group transformations.
7. The coordinates in terms of which all fields are defined are q-numbers of the form $X^{\mu} = x^{\mu} + iY^{\mu}(x)/M_c^2$ where $Y^{\mu}(x)$ is a free quantum gauge field, M_c is a constant with a value less than or equal to the Planck mass, and the x^{μ} are real-valued coordinates.
8. Covariant derivatives and Reality group representations of particle wave functions generate the fermion spectrum and gauge bosons embodying particle interactions.
9. The energy of the fundamental fermions and gauge bosons modulo interactions is real-valued.

10. Scalar particles, called Higgs bosons, generate particle masses through spontaneous breakdown.

This short list of axioms, admittedly based on complex theoretical constructs such as quantum field theory, leads to the New Standard Model. This can be seen in the detailed presentations in our book from Blaha (2011c) through Blaha (2013). We obtain the fermion mass spectrum including a Dark Matter sector, the symmetry group structure SU(3)⊗SU(2)⊗U(1)⊗SU(2)⊗U(1), the Standard Model interactions, and particle masses via the Higgs mechanism. The q-number coordinates cause all calculations in perturbation theory to be finite – no infinities.

In Blaha (2013) and earlier books we have hypothesized that the fundamental origin of mass, which after all are merely dimension-full constants in the Higgs lagrangian sector, is due to a larger theory in which the Higgs' dynamic equations generate mass terms as separation constants due to derivative terms in the coordinates in a sister universe. The sister universe can thus be viewed as "defining" mass, inertia, and the inertial reference frames of our universe. In addition the sister universe is responsible for four generations of fermions – three of which have been observed. See Blaha (2011c) pp. 282 - 284.

2. Sister Universe: Higgs Particles' Lagrangian Mass Terms; Origin of Mass, Time, and Inertial Reference Frames; Multiverse Clock; and Multiverse Observer

There's a hell of a good universe next door; let's go.
"One Times One" - E. E. Cummings

2.1 Reasons for a Sister Universe

In this book and earlier books we have suggested that there are weighty reasons to believe that a sister universe exists. In addition, new evidence is beginning to appear to support its existence. It appears that the sister universe is

1. The ultimate source of mass and inertia in our universe.
2. The "clock" that sets the time in our universe.
3. The source of fermion generations and mass mixing in our universe.
4. The "Observer" that makes a quantum multiverse of our universe.

We view the deep necessity of the sister universe to originate in the need for complex parallel physical processes. Section 1.3 describes the two principles of asynchronicity identifying the spatial principle as the source of the dimensionality of our universe. The temporal principle is the source of the dimensionality of the sister universe – also four dimensional.

Asynchronous Logic provides the equivalent of a clock for the synchronization of processes within large electrical systems such as VLSI chips. Similarly there is a need for a clock for a multiverse. For quantum gravity DeWitt[41] points out,

'"The variables ... [of the quantized Friedmann model] because of their lack of hermiticity, are not rigorously observable and hence cannot yield a measure of proper time

[41] DeWitt, B. S., Phys. Rev. **160**, 1113 (1987).

which is valid under all circumstances. ... . It is for this reason that we may say that "time" is only a phenomenological concept ... If the principle of general covariance is truly valid then the quantum mechanics of everyday usage with its dependence on the Schrödinger equations ... is only a phenomenological theory. For the only "time" which a covariant theory can admit is an intrinsic time defined by the contents of the universe itself. Any intrinsically defined time is necessarily non-Hermitean, which is equivalent to saying that there exists no clock, whether geometrical or material, which can yield a measure of time which is operationally valid under *all* circumstances, and hence there exists no operational method for determining the Schrödinger state function with arbitrarily high precision."

The lack of a clock within our universe invalidates quantum mechanics in principle. DeWitt concludes, "Thus [quantum gravity] will say nothing about time unless a clock to measure time is provided."

Unruh[42] also has an issue with the question of time:

"One of the key problems is that of time. We see and experience the world in terms of time. We see things grow, develop, and change. However, time does not enter into the Euclidean formulation of quantum gravity directly. In the usual Hamiltonian formulation, the Hamiltonian for quantum gravity is made up of densities which are the generators, not only of spatial coordinate transformations, but also of temporal coordinate transformations. The content of four of Einstein's equations, namely, the 6 „components, is that these generators are zero. Thus all wave functions are invariant under all spatial and all temporal coordinate transformations. There is nothing in the wave function or the amplitudes which refers to the coordinate t, or the corresponding points of the manifold in any way. How then do we recover the indubitable and ubiquitous experience we have of time? The standard answer is that our experience of time is actually an experience of different correlations between physical quantities in the world. Time is replaced by the readings of clocks. I know that time has changed, not through any direct experience with time, but because the hands of my watch have changed.

Although the implementation of this idea is actually extremely difficult in practice, and although I personally believe that one should formulate one's quantum theory of gravity so as to contain time explicitly, let us nevertheless pursue the consequences of this idea of time as defined internally, as the "reading" of a dynamic variable. For an observer inside the theory, his "time" is not the coordinate t. Rather his time is some one of the given dynamic variables of the theory: y or P. Thus although the coupling to the baby universes via the effective action S,. is independent of the coordinates t or x, that does not mean that the observer inside the theory will experience the interactions as being independent of time. For him and/or her, time is one of the dynamic variables and so it can depend on the various dynamic variables of the theory, even if it does not depend on the time coordinate t. In general one would expect the observer to see what looks to him like a time-dependent interaction with the baby universes. At one time, some

---

[42] Unruh, W. G., Phys. Rev. D **40**, 1053 (1989).

one of the baby universes may couple strongly to the large universe, while at some other time, another of the baby universes will couple more strongly."

We suggest the sister universe serves the role of a clock for our universe. And being a universe, it is an excellent clock. DeWitt points out, 'Because every clock has a "one-sided" energy spectrum, its ultimate accuracy must necessarily be inversely proportional to its rest mass. When the whole universe is cast in the role of a clock, the concept of time can of course be made fantastically accurate (at least in principle) ... ' Setting a mass scale using the sister universe, also sets[43] a time scale and resolves the issue of a clock for our universe. *In principle our sister universe validates the role of time in the Copenhagen interpretation of quantum mechanics.*

Thus the Higgs lagrangian sector mass, which is set by the sister universe, has numerous significant consequences. Section 1.10.5 summarizes the origin of the Higgs mass terms as separation constants between the two universes. Section 1.10.4 shows how the Higgs fields create the fermion generations, and generation mass mixing terms, in our universe. Again a separable dynamic equation straddling the two universes is the source.

Attempts to create a quantum gravity theory have to confront the need for an *observer* in any quantum theory in the Copenhagen interpretation. DeWitt points out,

"The Copenhagen view depends on the assumed a priori existence of a classical level to which all questions of observation may ultimately be referred. Here, however, the whole universe is the object of inspection; there is no classical vantage point, and hence the interpretation question must be re-argued from the beginning. While we do not wish to stress this point unduly, since, after all, the Friedmann model ignores the vast complexities of the real universe, it is nevertheless clear that the quantum theory of space-time must ultimately force a deviation from the traditional Copenhagen doctrine."

And Unruh states

"One of the key features in the interpretation of such transition amplitudes, or wave functions, is the idea that we, as observers are also a part of the Universe as a whole. We, as physical observers, must be describable from within the theory and not as observers external to

---

[43] For example the Planck time value is set by the Planck mass.

the theory as in usual quantum mechanics. In usual quantum mechanics, the interpretation is usually given in terms of observers that are outside of the theory. There one makes a split, with the quantum world at one side of the split, and the observer on the other. von Neumann argued that the predictions of quantum mechanics, at least under certain assumptions, are independent of the exact location of that split, but Bohr argued adamantly for the necessity of such a split (classical observers and quantum world). *There is a great difficulty in setting up such a split for physical observers contained within and influenced by a quantum universe,* [itallics added] and for the Universe as a whole, especially including gravity, one cannot argue that the predictions will be independent of where one puts the split. Since all energies interact gravitationally, and our observations are surely energetic phenomenon, the treatment of the energetics of observation as classical would lead to different predictions than if they were treated quantum mechanically. One is therefore forced to devise an interpretation of quantum mechanics in which the observer is part of the quantum system, rather than outside the quantum system.

This means that the interpretation of these transition amplitudes becomes somewhat non-intuitive. One must ask what the system looks like from within, from the viewpoint of an observer who is part of that world, rather than being able to interpret them directly in terms of probabilities for observations made by an external observer."

While the *observer* question is addressed by a number of authors, the proposed answers are not entirely convincing. *The existence of a sister universe provides a macroscopic quantum observer for our universe.* And our universe provides a macroscopic quantum observer for our sister universe. Thus the quantum observer issue is resolved by the sister universe.

These considerations lead us to view the sister universe as a critical solution to the above problems.

By establishing Asynchronous Logic principles as the basis for the existence of two universes and for setting the number of dimensions in each universe we have found deeper principles of organization for the foundations of physics. These principles serve to enable the coordination of complex physical processes.

Usually we look at particle processes primarily from a space-time perspective: particles collide and produce new particles. We primarily think of the incoming and outgoing particles in a collision. However, considering the set of fundamental particles – and the particle transforming interactions in themselves – neglecting space-time and momentum considerations – leads us to view particles as constituting an alphabet and their interactions as a type of computer grammar.[44] Then the Asynchronicity Principles enable us to bring in space-time in a way that gives us the maximum complexity with the most minimal assumptions. As Leibniz[45] points out our universe has maximal complexity with minimal assumptions.

2.2 General Features of the Sister Universe

The sister universe has a set of features based on the assumption that it resembles our universe to a maximal extent dynamically but is likely to have a different amalgam and distribution of matter in planets, stars, galaxies and super-galaxies due to the vagaries of chance during the course of its evolution. In this section we will list features that we should expect and briefly describe some of them.

1. The sister universe has four dimensions, is curved and closed, a metric tensor that in the flat space-time approximation is similar to the metric of our universe, complex Lorentz group covariance in the flat space-time approximation, the same Reality group as our universe, ... In short the same space-time features as our universe.
2. The fermions and bosons are the same as in our universe but not necessarily with the same masses (as we will see in the following section in the case of fermions.)
3. The dynamic equations of the fermions and bosons may have different coupling constants and masses.
4. The sister universe has the same second quantization procedures, the same Faddeev-Popov mechanism, and the same Two-Tier mechanism to make all field theory calculations finite (no infinities).
5. The sister universe has the same form of classical and quantum general relativity as our universe.
6. The dynamic equations of both universes are combined in a separable form. Consequently the universes form a direct product mathematically.
7. The sister universe is a surface in the Flatverse parameterized by eight complex Flatverse coordinates (or 16 real-valued coordinates). The 8

[4444] This conceptual approach was first described in Blaha (1998) who went on to characterize our universe as one enormous word evolving in time.

[45] See Rescher (1967).

sister universe complex Flatverse coordinates plus the 8 complex Flatverse coordinates of our universe together comprise the 16 complex coordinates of the Flatverse. The separation of the Flatverse coordinates into a set for our universe and a set for the sister universe can be changed by performing a rotation of the Flatverse coordinates. The resulting additional complexity does not appear to have any advantages.

8. The two universes do not overlap and may not have a common surface. There may be "space between the spaces." Or a wormhole connection. Together they do not occupy the entire Flatverse because of its infinite extent.
9. The sister universe, like our universe, has total charge, and all other additive quantum numbers, zero. The total energy is zero as well with the energy of matter cancelled by the gravitational field energy. Thus there is no gravitational attraction between the universes.
10. The sister universe metric may differ from that of our universe. The sister universe may or may not be expanding.
11. The sister universe may have a different topological structure than our universe.

This list of sister universe properties is based on the assumption that the sister universe is similar to our universe. This assumption is sensible in view of the study of our universe in this book and preceding books by the author that show the form and particle spectrum of the Standard Model can be directly based on geometry. Item 1 above implies the sister universe must have the same form and particle spectrum as the Standard Model in our universe.

2.3 Embedding the Two Universes in the Flatverse/Multiverse

Our two universe theory is the first example of a multiverse with some experimental support by its explication of the Higgs mass terms and its explanation of the origin of fermion generations and mass mixing. In this section we will discuss the role of the Flatverse as the stage upon which the multiverse performs.

Earlier we pointed out how the existence of a sister universe resolves problematic features of the multiverse and quantum gravity such as the observer and clock time issues. We shall assume the Flatverse is the multiverse. In the next chapter we will describe a new theory of the multiverse with a generalization of the Wheeler-DeWitt equation to a complex Euclidean space – the Flatverse, and to the case of complex Lorentzian space-times. These features change the hyperbolic Riemannian manifold of 3-geometries from six to nine dimensions. They also change the features of the superspace infinite-dimensional manifold as well as changing other features of quantum gravity and the multiverse.

In the previous chapter we discussed embedding our *complex* universe in a 16-dimensional flat space using

$$z_i = f_i(x) \tag{1.4}$$

where x is a complex 4-vector in our universe[46] (as in chapters 2 and 3 of Blaha (2012b)), and where z_i are the coordinates (for i = 1, ..., 16) of a flat space point. The functions f_i map our universe into the 16 dimensional real, flat space as a complex 4-dimensional surface (an 8-dimensional real-valued surface.)

We then showed that the 16-dimensional real space must be replaced with a 16-dimensional *complex* flat space that we have called the Flatverse in Blaha (2012b) with invariant interval

$$ds^2 = g_{ij}dz^i dz^j \tag{1.5}$$

with the metric satisfying

$$g_{ij} = g_{ji} \tag{1.6}$$

The metric tensor $g_{\mu\nu}$ of the universe is

$$g_{\mu\nu} = \partial f_j/\partial x^\mu \, \partial f_j/\partial x^\nu \tag{1.7}$$

with an implied sum over the subscript j.[47] Outside of closed universes, which we assume are the only aggregations of mass-energy, the Flatverse is truly flat with a Euclidean metric g_{ij} = 1 for all i = j and zero otherwise.

---

[46] Our universe is then a complex 4-surface within the 16-dimensional complex flat space.

[47] Recall that the complex four-dimensional metric has 16 independent components thus leading to an 8-dimensional complex embedding in the 16-dimensional complex Flatverse.

Our universe can be contained within eight of the complex dimensions of the Flatverse:

$$z_j = f_j(x) \tag{2.1}$$

for j = 1, 2, ... , 8. The remaining eight complex dimensions y can contain a sister universe which we posit is similar in most respects to our universe:[48]

$$z_j = f_j(y) \tag{2.1a}$$

for j = 9, 10, ... , 16.

These two universes do not fill the entire Flatverse since it is of infinite extent, and they are closed surfaces. Thus there could be other universes in the Flatverse. If there is a baryonic quantum gauge field in the Flatverse that can spawn universes via quantum fluctuations of the Flatverse universe, then other universes are a distinct possibility. We discuss this feature in more detail in chapter 4.

Other questions that arise are whether the two universes overlap, or have a common surface part. Are they "adjacent" to each other? If they overlap then their respective metrics would have an interrelation that is not seen and that would conflict with general relativity.

Whether they are adjacent is an open question. This question raises the issue of wormholes between the universes as well as the impact on adjacent universes of boundary conditions on any common surface. Experimentally, we have not seen a wormhole.[49] But space is very large and we have many unexplored and poorly understood regions of space. So the questions remain open. We will explore quantum universes and wormholes theoretically in our complex version of the Wheeler-DeWitt equation in chapter 3 together with the impact of the Reality group on the Wheeler-DeWitt formalism.

---

[48] A four complex dimension Lorentzian space-time.

[49] The Great Attractor found in cosmological observation does seem to have a curvature due to unseen causes may be a highly curved part of the universe around a wormhole.

2.4 Separation of Particles' Dynamic Equations between the Two Universes

This chapter[50] describes how separable dynamic equations can be formed for all particles. The equations separate into a piece for our universe and a piece for the sister universe. The definition of lagrangians and other quantum field theory constructs can be done either prior to separation for both universes or constructed separately for each universe. Two-Tier quantization prevents the appearance of infinities in four dimensional theories and in higher dimensional theories as well.

This section shows how the Higgs Mechanism generates fermion and gauge boson masses as well. In Blaha (2011c), which appeared prior to the experimental discovery of Higgs particles, we showed how the separation of dynamic equations used mass terms as separation constants providing an alternative to the Higgs Mechanism. In this section we show that the separation constants' masses introduce a new degree of freedom in particle mass values that gives flexibility to the determination of masses via the Higgs Mechanism. Masses may arise due to a combination of separation constant masses and Higgs generated masses.[51]

2.4.1 Separable Multi-Universe Higgs Dynamic Equations

In Blaha (2013) we pointed out that the Higgs sector of the Standard Model lagrangian had a Higgs mass term with the "wrong" sign in the sense that it made each Higgs particle a tachyon. The tachyonic nature was submerged due to the quartic interaction term which together with the mass term produced a spontaneous breakdown.

We consider an example of the creation of Higgs particle mass terms for the case of complex scalar particle fields.[52] The realistic case of multiplets of Higgs particles is a direct generalization. The Higgs sector lagrangian density terms have the form

---

[50] Much of the material in this section appears in Blaha (2011c) together with additional comments.

[51] It is possible, but unlikely in this author's view, that the separation constants' masses are all zero.

[52] Chapter 1 of Blaha (2012b) contains this discussion and example, which we place here for the sake of completeness.

$$\mathcal{L}(\phi(x)) = \partial^{\mu}\phi\partial_{\mu}\phi^* - V(\phi, \phi^*) \tag{2.3}$$

where

$$V(\phi, \phi^*) = m^2\phi\phi^* + \beta(\phi\phi^*)^2 \tag{2.4}$$

If $\beta > 0$ and $m^2 < 0$ spontaneous symmetry breaking occurs and the ϕ field acquires a constant part ϕ_0

$$\phi(x) = \phi_0 + \delta\phi(x) \tag{2.5}$$

with

$$\phi_0 = -m/\sqrt{\beta} \tag{2.6}$$

up to a phase factor. Note that a negative value for m^2 indicates the ϕ field is a tachyon field. The tachyonic nature is transmuted to be a constant throughout space and time by the quartic interaction term.

We noted further that β is a dimensionless constant. Consequently the mass term is the only "dimensioned" constant in the lagrangian and the resulting dynamical equations.

We will now show that the mass term can originate through a separation of equations IF the true lagrangian is the sum of pieces from the two universes. We called this the Dimensional Mass Mechanism. (See Blaha (2011c).)

We assume n Higgs bosons that are functions of our space-time coordinates labeled x and four space-time coordinates labeled y in the sister universe.[53]

The familiar Higgs boson field equations have a quartic interaction if we temporarily ignore the sister universe and its mass term contribution. We define

$$B^a(x) = \Box\phi^a + \beta_a\phi^{a4} = 0 \tag{2.7}$$

Following a parallel development to the fermion case of Blaha (2011c) we consider a coupled universes' generalization of eq. 2.7:

$$B^a(x, y) \equiv B_U{}^a(x)\phi_S{}^a(y) + \phi_U{}^a(x)B_S{}^a(y) = 0 \tag{2.8}$$

---

[53] See Blaha (2011c) for a more detailed discussion of the "other" space-time.

where U signifies our universe and S signifies the sister universe.

This equation results from the lagrangian density term

$$V_a(x, y)\, B^a(x, y) \tag{2.9}$$

where $V_a(x, y)$ is a subsidiary field. Variation with respect to $V_a(x, y)$ yields eq. 2.8. Variation with respect to $\phi_U{}^a(x)$ and $\phi_S{}^a(y)$ yield the other dynamical equations. The linearity in $V_a(x, y)$ in these dynamical equations imply $V_a(x, y)$ is separable and the product of two boson fields:

$$V_a(x, y) = V_{Ua}(x)V_{Sa}(y) \tag{2.10}$$

Dividing eq. 2.8 by $\phi_U{}^a(x)\phi_S{}^a(y)$ yields a separable equation with the result:

$$B_U{}^a(x)\,/\phi_U{}^a(x) = -m_a{}^2 \tag{2.11}$$

and

$$B_S{}^a(y)_S/\phi_S{}^a(y) = m_a{}^2 \tag{2.12}$$

where the right side of each equation is a constant mass squared (required on dimensional grounds).

Consequently a Higgs boson field equation in our universe acquires mass terms

$$\Box\phi_U{}^a + m_a{}^2\phi_U{}^{a\,2} + \beta_a\phi_U{}^{a\,4} = 0 \tag{2.13}$$

and thus provides an origin for the Higgs mass term based on the separation of variables.

Interestingly the mass terms in the sister universe have the opposite sign and thus do not cause spontaneous breakdown:

$$\Box\phi_S{}^a + m_a{}^2\phi_S{}^{a\,2} + \beta_a\phi_S{}^{a\,4} = 0 \tag{2.14}$$

with the result that fermions and vector bosons do NOT acquire masses via the Higgs Mechanism although they may have masses due to bare mass terms and their renormalization in quantum field theory. *The possibility of a sister universe in which all particles are massless (except the Higgs particles) is remarkable.*

Massless QED[54] in our universe was essentially solved up to an eigenvalue condition (that this author[55] calculated approximately to infinite order but without success in obtaining the correct value of the fine structure constant.) The complexity of The Standard Model, even if massless, appears to make a "solution" impossible due to the complexity of non-abelian gauge boson terms and the need for Faddeev-Popov ghosts.

The existence of the sister universe is the necessary price for a deeper understanding the origin of mass. A Higgs particle is partly in our universe and partly in a sister universe. Mass terms arise as separation constants when the parts of dynamic equations are separated for each universe. The mechanism is analogous to the appearance of a mass in the Schwarzschild solution in General Relativity.

Since the Higgs mass term sets the scale for all masses in The Standard Model and in General Relativity we can attribute the origin of not only the mass scale but also of inertia, and inertial reference frames to the existence of a sister universe (which replaces the "fixed stars" of Mach). Together with other important roles of the sister universe, such as the derivation of fermion generations we conclude there is significant theoretical support for at least two universes in the multiverse. Fermion generations and Higgs masses are the first partial evidence for a multiverse. Until now the multiverse was speculative.

2.4.2 Separable Multi-Universe Fermion Dynamic Equations

The sister universe was shown to be the likely origin of the four generations of each fermion species in our universe in section 1.10.5, and of the Higgs particles mass terms and mass mixing in section 1.10.5 and the preceding section and books. In section 1.10.5 we found we could factorize *each fermion species* dynamic equation spanning both universes:

---

[54] The Johnson-Baker-Willey model: see M. Baker and K. Johnson, Phys. Rev. **D8**, 1110 (1973) and references therein.

[55] S. Blaha, Phys. Rev. **D9**, 2246 (1974). Although this solution summed an infinite number of diagrams' contributions – more than any other calculation in QED known to this author, and although the solution agreed with the exact calculated values to 4th order in α, an essential singularity (predicted by S. Adler) was not found (although a complex analytic structure was found). The first zero was at $\alpha = 1$. For this reason the Johnson-Baker-Willey program for creating a finite QED seems to have failed.

$$[D_U \otimes I_S + I_U \otimes D_S + g_S\Phi(x, y)M]\psi_R(x, y) = 0 \qquad (1.24)$$

where U represents our universe, S represents the sister universe, and where $\Phi(x, y)$ is a Higgs field in both universes as explained in Blaha (2013). I is the identity matrix. Explicitly, with indices displayed, and the D terms expanded according to each of the four species, eq. 1.24 is[56]

$$[\gamma^{\nu}{}_{ab}\mathcal{D}_{\nu}(x) + \gamma^{\nu}{}_{cd}\mathcal{D}'_{\nu}(y) + g_S\Phi(x, y)M_{cd}\,\delta_{ab}]\psi_{Rbd}(x, y) = 0 \qquad (1.25)$$

Upon spontaneous breakdown the Higgs field becomes a constant (plus an additional variable Higgs field part that is not displayed) which we normalize to unity

$$g_S\Phi(x, y) \rightarrow g_S \qquad (1.26)$$

We separate the solution into parts for each universe

$$\psi_{Rbd}(x, y) = \psi_{Rbd}(x)\psi'_{Rd}(y) \qquad (1.27)$$

and then separate the equations:[57]

$$\gamma'^{\mu}{}_{cd}\mathcal{D}'_{\mu}(y)\psi'_{Rd}(y) = 0 \qquad (1.28)$$

$$(\gamma^{\nu}{}_{ab}\mathcal{D}_{\nu}(x)\delta_{cd} + g_S M_{cd}\delta_{ab})\psi_{Rbd}(x) = 0 \qquad (1.29)$$

or, in matrix notation, eq. 1.29 – the equation for the fermion generations – in our universe becomes

$$(\gamma^{\nu}\mathcal{D}_{\nu}(x) + g_S M)\psi_R(x) = 0 \qquad (1.30)$$

where $\psi_R(x)$ is four generations of a fermion species right-handed spinors. Thus the fermion generations emerge from geometry – the impact of the sister universe on our universe.

The fermion equation eq. 1.28 for a species in the sister universe could also have a bare mass term if such a term had been inserted in eq. 1.24:

---

[56] The prime ' such as that on $\mathcal{D}'$ indicates the sister universe.

[57] This separation enables the fermions of the sister universe to have a zero mass (or one set by a bare mass term, as opposed to a mass matrix that is not a multiple of the identity matrix in its Dirac-like equation.

$$[D_U \otimes I_S + I_U \otimes D_S + m_0 I_U \otimes I_S + g_S \Phi(x, y)M]\psi_R(x, y) = 0 \qquad (1.24a)$$
$$[\gamma'^{\mu}{}_{cd}\mathcal{D}'_{\mu}(y) + m_0 I_U \otimes I_S]\psi'_{Rd}(y) = 0 \qquad (1.28a)$$

Eq. 1.28a shows that each fermion species in the sister universe has only one generation. This difference is an important differentiating feature of the sister universe from our universe since it also excludes the mass mixing and consequent pattern of decays that we find in our universe.

There is another possible source for fermion masses – as separation constants for eq. 1.25 after implementing the Higgs mechanism. The appearance of separation constants[58] would modify eqs. 1.28 and 1.29 to

$$[\gamma'^{\mu}{}_{cd}\mathcal{D}'_{\mu}(y) - M_{sep}\delta_{cd}]\psi'_{Rd}(y) = 0 \qquad (1.28a)$$
$$[\gamma^{\nu}{}_{ab}\mathcal{D}_{\nu}(x)\delta_{cd} + g_S M_{cd}\delta_{ab} + M_{sep}\delta_{cd}]\psi_{Rbd}(x) = 0 \qquad (1.29b)$$

thus endowing fermions in the sister universe with masses. This possibility appears more physically acceptable than having only massless fermions in the sister universe. It is also logically consistent with the separation constants appearing in the Higgs particle case described in the previous section.

On the negative side, fermion masses in our universe would then not be solely due to the Higgs mechanism. This extra degree of freedom in the masses would change the numerics of the Higgs mass determinations. For example, the plus signed third term in eq. 1.29b would then have to have its value compensated by the Higgs mechanism term.

2.4.3 Separable Multi-Universe Gauge Boson Dynamic Equations

We *could* choose to use massive vector bosons *ab initio* and obtain a *fully renormalizable* (indeed, finite) ElectroWeak Theory by using the Two-Tier quantization procedure (Blaha (2003) and (2005a) – Part 3). A massive vector boson lagrangian would then have the form:

$$\mathcal{L}_{Y\text{-}M} = -\tfrac{1}{4}\, F^{a\mu\nu}(x)F^{a}{}_{\mu\nu}(x) + \tfrac{1}{2}\, m^2 A^{a\mu}A^{a}{}_{\mu} \qquad (2.15)$$

where

---

[58] As in Blaha (2011c).

$$F^a_{\mu\nu}(x) = D_\nu(x)A^a_\mu(x) - D_\mu(x)A^a_\nu(x) + gf^{abc}A^b_\mu(x)A^c_\nu(x) \qquad (2.16)$$

The form of the differential operator D_ν depends on whether the Yang-Mills field is "normal" or complexon. For a normal Yang-Mills Field $D_\nu = \partial_\nu$ while for a complexon Yang-Mills field D_ν is specified by an equation of Blaha (2011c):

$$D_0 = \partial/\partial x^0$$
$$D_k = \partial/\partial x_r^k + i\,\partial/\partial x_i^k \qquad (17.59)$$

The constants f^{abc} are the algebra's structure constants.

However we will now consider mechanisms for dynamic mass generation. Prior to the discovery of the Higgs particle we suggested mass generation as a separation constant in gauge boson fields dynamic equations based on our view that experiment both leads theory and confirms it. We begin therefore with the separation mass mechanism of Blaha (2011c).

Consider the separation of non-abelian gauge boson dynamic equations for the two universes described in Blaha (2011c). (Similar considerations apply for abelian gauge boson dynamic equations.) The dynamical field equation of a conventional 4-dimensional massless Yang-Mills field is

$$G^{a\nu}(x) = (D_\mu(x)\delta^{ac} + gf^{abc}A^b_\mu(x))F^{c\mu\nu}(x) = 0 \qquad (2.17)$$

where $F^{a\mu\nu}(x)$ is

$$F^a_{\mu\nu}(x) = D_\nu(x)A^a_\mu(x) - D_\mu(x)A^a_\nu(x) + gf^{abc}A^b_\mu(x)A^c_\nu(x) \qquad (2.18)$$

The form of the differential operator D_ν depends on whether the Yang-Mills field is "normal" or complexon. For a normal Yang-Mills Field $D_\nu = \partial_\nu$ while for a complexon Yang-Mills field D_ν is specified by eq. 17.59 above.

Following a parallel development to the fermion case we consider a two universe generalization of eq. 2.17:[59]

[59] The form is determined by requiring the equation be second order, and that the factors in both universes are vectors.

$$G_U{}^{a\nu}(x)A_S{}^{b\mu}(y) + A_U{}^{a\nu}(x)G_S{}^{b\mu}(y) = 0 \tag{2.19}$$

where U and S represent our universe and its sister universe respectively.

This equation follows from the lagrangian density terms

$$V_{a\nu b\mu}(x, y)[G_U{}^{a\nu}(x)A_S{}^{b\mu}(y) + A_U{}^{a\nu}(x)G_S{}^{b\mu}(y)] \tag{2.20}$$

where $V_{a\nu b\mu}(x, y)$ is a subsidiary field. Variation with respect to $V_{a\nu b\mu}(x, y)$ yields eq. 2.19. Variation with respect to $A_U{}^{a\nu}(x)$ and $A_S{}^{b\mu}(y)$ yield the other dynamical equations for this subsector. The linearity in $V_{a\nu b\mu}(x, y)$ in these equations implies $V_{a\nu b\mu}(x, y)$ is separable and the product of two gauge fields:

$$V_{a\nu b\mu}(x, y) = V_{1a\nu}(x)V_{2b\mu}(y) \tag{2.21}$$

Dividing eq. 2.19 by $A_S{}^{b\mu}(y)A_U{}^{a\nu}(x)$ yields a separable equation with the results:

$$G_U{}^{a\nu}(x)/A_U{}^{a\nu}(x) = -m^{a2} \tag{2.22}$$

and

$$G_S{}^{a\nu}(y)/A_S{}^{a\nu}(y) = m^{a2} \tag{2.23}$$

where the right side of each equation is a constant mass squared (required on dimensional grounds). Eqs. 2.19 are not covariant under local gauge transformations. This conclusion is consistent with the acquisition of mass by the fields (eqs. 2.22 and 2.23) breaking gauge symmetry.

Consequently the U Yang-Mills fields become massive Yang-Mills fields and satisfy

$$(D_\mu(x)\delta^{ac} + gf^{abc}A_U{}^b{}_\mu(x))F_U{}^{c\mu\nu}(x) + m^{a2}A_U{}^{a\nu}(x) = 0 \tag{2.24}$$

and the S Yang-Mills fields satisfy

$$(D_\mu(y)\delta^{ac} + gf^{abc}A_S{}^b{}_\mu(y))F_S{}^{c\mu\nu}(y) - m^{a2}A_S{}^{a\nu}(y) = 0 \tag{2.25}$$

Thus we have an alternate mechanism to generate masses for Yang-Mills fields. The theories embodying these fields are fully renormalizable (as noted earlier) using Two-Tier quantization. Eq. 2.25 has a negative mass term that would make the gauge fields tachyonic. The Higgs mechanism can remove this apparent flaw.

The Higgs Mechanism can be used together with the above separation of equations approach or independently to obtain masses for non-abelian gauge fields. To achieve a combined approach we modify eq. 2.17 to

$$G_H{}^{a\nu}(x) = (D_\mu(x)\delta^{ac} + gf^{abc}A^b{}_\mu(x))F^{c\mu\nu}(x) + A^{a\nu}\phi^{a2}(x) + \ldots = 0 \qquad (2.17a)$$

exposing the Higgs term that becomes a mass term. The … indicates all remaining interaction terms for the gauge field plus any counter terms necessary to keep the electromagnetic field massless.

Following the same procedure as above with the Higgs field becoming a mass term through spontaneous breakdown: $\phi^{a2}(x) \rightarrow M^{a2}$ we arrive at the Higgsian plus separation constant equation for our universe

$$(D_\mu(x)\delta^{ac} + gf^{abc}A_U{}^b{}_\mu(x))F_U{}^{c\mu\nu}(x) + m^{a2}A_U{}^{a\nu}(x) + M^{a2}A_U{}^{a\nu}(x) = 0 \qquad (2.24a)$$

and for the sister universe

$$(D_\mu(y)\delta^{ac} + gf^{abc}A_S{}^b{}_\mu(y))F_S{}^{c\mu\nu}(y) - m^{a2}A_S{}^{a\nu}(y) + M^{a2}A_S{}^{a\nu}(y) = 0 \qquad (2.25a)$$

Thus $M^{a2} - m^{a2}$ must be positive to avoid tachyonic behavior for sister universe gauge fields.

It remains for experiment and further theoretical work to determine whether masses derived in part from separation constants are required.

2.4.4 Separable Multi-Universe Gravitational Dynamic Equations

The force of gravitation has its source in mass and energy. Since there is no mass-energy in the Flatverse (which we take to be the multiverse), we assume that gravitation is non-existent outside of universes. Put another way, the Flatverse is flat outside of universes. The Flatverse is assumed to have a Euclidean metric $g_{ij} = \delta_{ij}$, the Kronecker delta for all i and j. The invariant interval is

$$ds^2 = g_{ij}dz^i dz^j \qquad (1.5)$$

Within a universe the metric g_{ij} satisfies

$$g_{ij} = g_{ji} \tag{1.6}$$

and is given by

$$g_{\mu\nu} = \partial f_j/\partial x^\mu \, \partial f_j/\partial x^\nu \tag{1.7}$$

where the complex Flatverse coordinates z_i are

$$z_i = f_i(x) \tag{2.1}$$

for i =1, 2, ... , 16 and x represents the coordinates parameterizing universes.

We have shown that our universe, and the sister universe, are each parameterized by four complex-valued coordinates. Further we have shown that our universe can be embedded in an eight complex coordinates region of the Flatverse with z_i = 1, 2, ..., 8; and the sister universe similarly can be embedded in a different eight complex coordinates region of the Flatverse with z_i = 9, 10, ..., 16 with no overlap of universes although the possibility exists of a common surface. A rotation of the Flatverse coordinates could of course change the Flatverse coordinates corresponding to each universe. We will choose to use the clean separation of coordinates given above for the sake of simplicity.

Given this picture we can define a dynamical equation for our universe and our sister universe which is separable fundamentally because of the independence of the Flatverse coordinates of each universe (by the discussion in the paragraph following eq. 2.1 above. The dynamics equation is

$$\begin{aligned}&[R_{U\mu\nu}(x) - g_{U\mu\nu}(x)R_U(x) + \lambda_U g_{U\mu\nu}(x) + 8\pi G T_{U\mu\nu}(x)]g_{S\alpha\beta}(y) + \\ &+ g_{U\mu\nu}(x)[R_{S\alpha\beta}(y) - g_{S\alpha\beta}(y)R_S(y) + \lambda_S g_{S\alpha\beta}(y) + 8\pi G T_{S\alpha\beta}(y)] = 0\end{aligned} \tag{2.26}$$

using the familiar tensors of General Relativity where λ_U and λ_S are cosmological constant terms in our universe and our sister universe. The separation of equations is evident and symbolically

$$\begin{aligned}&[R_{U\mu\nu}(x) - g_{U\mu\nu}(x)R_U(x) + \lambda_U g_{U\mu\nu}(x) + 8\pi G T_{U\mu\nu}(x)]/g_{U\mu\nu}(x) = -\lambda_C \\ &[R_{S\alpha\beta}(y) - g_{S\alpha\beta}(y)R_S(y) + \lambda_S g_{S\alpha\beta}(y) + 8\pi G T_{S\alpha\beta}(y)]/g_{U\mu\nu}(y) = \lambda_C\end{aligned} \tag{2.27}$$

or

$$[R_{U\mu\nu}(x) - g_{U\mu\nu}(x)R_U(x) + (\lambda_U + \lambda_C)g_{U\mu\nu}(x) + 8\pi G T_{U\mu\nu}(x)]/g_{U\mu\nu}(x) = 0 \quad (2.28)$$
$$[R_{S\alpha\beta}(y) - g_{S\alpha\beta}(y)R_S(y) + (\lambda_S - \lambda_C)\lambda_S g_{S\alpha\beta}(y) + 8\pi G T_{S\alpha\beta}(y)]/g_{U\mu\nu}(y) = 0$$

If we wish a similar qualitative expansion of the sister universe $\lambda_S - \lambda_C$ must at least be positive. To obtain the same expansion rate then we must have $\lambda_S = \lambda_U$ and $\lambda_C = 0$. The question of the expansion rate of the sister universe is an open one.

2.4.4.1 A Wormhole Connection between the Two Universes?

Having two universes it is reasonable to ask if they are in any way connected. Clearly they are disjoint except for a possible common piece of surface. But can they be connected by a wormhole? It appears that the answer to this question is possibly yes. As Hawking[60] points out in a particular case there can be a wormhole connection. It could be with a wormhole whose size approaches zero.[61] *Thus a physical wormhole connection between the universes is possible.*

2.4.5 Separable Multi-Universe Quantum Coordinate Field Y^μ

The quantum coordinates of our universe (section 1.2) are easily specified if our universe is flat. However our universe (and most likely its sister universe) are curved and closed 4-dimensional complex space-times. In the general case where our universe is curved the complex-valued Flatverse coordinates satisfy

$$z_i = f_i(X(z)) \quad (2.1b)$$

for i =1, 2, ... , 8 and X(z) are the 4-dimensional q-number coordinates parameterizing our universe. The solution of eq. 2.1a *if* our universe is flat is

$$X^\mu(z) = z^\mu + i\, Y^\mu(z)/M_c^2 \quad (1.35)$$

[60] Hawking, S. W., Phys. Rev. **D 37**, 904.

[61] In Hawking's notation on p. 905 as the wormhole radius b → 0.

where M_c is a very large mass scale of perhaps the order of the Planck mass, where the index $\mu = 1, 2, 3, 4$ ($X^4(z) = X^0(z)$ is the time coordinate), and where $Y^\mu(z)$ is a free QED-like abelian gauge field whose argument is a 4-dimensional vector z^μ.

If our universe is just slightly curved, as it is, then eq. 1.35 is still a good approximation. Suppose the function $f_i(y)$ is defined implicitly by

$$f_i(y) = y_i - iY_i(f(y))/M_c^2 \tag{2.29}$$

where f(y) represents the first four functions $f_k(y)$ for k = 1, 2, 3, 4. Then

$$\begin{aligned} f_i(X(z)) &= z_i + i\,Y_i(z)/M_c^2 - iY_i(f(X(y))/M_c^2 \\ &= z_i + \mathcal{O}(M_c^{-4}) \end{aligned}$$

or

$$z_i \cong f_i(X(z)) \tag{2.30}$$

for i = 1, 2, 3, 4. Eq. 2.30 is thus correct to order $\mathcal{O}(M_c^{-4})$ where M_c is of the order of, or less than, the Planck mass. The same considerations apply to the sister universe.

The $Y^\mu(z)$ field, a free abelian 4-dimensional gauge field, has the field strength

$$F_{\mu\nu}(x) = D_\nu(x)Y_\mu(x) - D_\mu(x)Y_\nu(x) \tag{2.31}$$

and the dynamical equation

$$G^\nu(x) = D_\mu(x)F^{\mu\nu}(x) = 0 \tag{2.32}$$

where the differential operator $D_\nu = \partial_\nu$.

The separable dynamic equation for $Y^\mu(z)$, a free abelian 4-dimensional gauge field, in each of the two universes is

$$G_U{}^\nu(x)Y_S{}^\mu(y) + Y_U{}^\nu(x)G_S{}^\mu(y) = 0 \tag{2.33}$$

where U and S represent our universe and its sister universe respectively.

This equation follows from the lagrangian density terms

$$V_{\nu\mu}(x, y)[G_U{}^\nu(x)Y_S{}^\mu(y) + Y_U{}^\nu(x)G_S{}^\mu(y)] \tag{2.34}$$

where $V_{\nu\mu}(x, y)$ is a subsidiary field. Variation with respect to $V_{\nu\mu}(x, y)$ yields eq. 2.33. Variation with respect to $Y_U^{\nu}(x)$ and $Y_S^{\mu}(y)$ yield the other dynamical equations for this subsector. The linearity in $V_{\nu\mu}(x, y)$ in these equations implies $V_{\nu\mu}(x, y)$ is separable and the product of two gauge fields:

$$V_{\nu\mu}(x, y) = V_{1\nu}(x)V_{2\mu}(y) \qquad (2.35)$$

Dividing eq. 2.33 by $Y_S^{\mu}(y)Y_U^{\nu}(x)$ yields a separable equation with the results:

$$G_U^{\nu}(x)/Y_U^{\nu}(x) = -m^2 \qquad (2.36)$$

and

$$G_S^{\nu}(y)/Y_S^{\nu}(y) = m^2 \qquad (2.37)$$

where the right side of each equation is a constant mass squared (required on dimensional grounds). Since these fields are massless in our New Standard Model we set $m^2 = 0$.

Thus we have the $Y^{\nu}(x)$ fields for Two Tier renormalization in both universes.

2.4.6 From Asynchronous Logic to Dynamical Equations to Lagrangians

The derivation of our fundamental theory of elementary particle physics begins with Asynchronous Logic (a 4-valued logic) requiring a four dimensional matrix that in turn leads to a 4-dimensional space-time and complex Lorentz group symmetry based on a Lorentzian invariant interval. From complex Lorentz group boosts we determine that four species of fermions are required. The need for a Reality group to transform complex coordinates into real-valued coordinates forces us to define covariant derivatives for use in dynamical fermion equations. Only at this point does the definition of lagrangians and hamiltonians which summarize the interacting dynamical equations and enable second quantization become feasible.

Thus Lagrangians and Hamiltonians are secondary constructs in this derivation. Consequently elementary particle and gravitational theory is a universe by universe construction. The preceding subsection discussions reflect this approach. *Lagrangians for dynamical equations are to be defined after the separation of the dynamical equations into a part for our universe and a part for the sister universe.*

2.5 Sister Universe Symmetries and Particle Spectrum

This section summarizes the major features of our sister universe to the extent that we can determine them from a similar logical derivation and from the separated dynamical equations of fermions and bosons. We will primarily point out the ***differences*** between our universes and the sister universe to avoid undue repetition.

2.5.1 Space-Time Symmetries

The sister universe has a complex 4-dimensional space-time with the Lorentzian invariant interval and the complex Lorentz group as its symmetry group. Complex coordinates are transformed into real-valued coordinates using SU(3)⊗SU(2)⊗U(1)⊗SU(2)⊗U(1) Reality group transformations.

2.5.2 Particle Symmetries and Interactions

Since the Reality group of our universe is determined by space-time geometry the New Standard Model symmetry group of the sister universe is the same as that of our universe. Since tachyons appear in sister universe ElectroWeak doublets the sister universes New Standard Model also has Parity Violation. The sister universe has quark confinement just as our universe.

Thus the sister universe New Standard Model has the same form as our universe except that there is only one generation of fermions.

The values of coupling constants, particle masses, and symmetry breaking parameters in the sister universe may, or may not, be the same as their values in our universe.

2.5.3 Fundamental Particle Spectrum

The form of the fermion and boson particle spectrum in the sister is the same as the form of the particle spectrum in our universe with three possible differences: the values of particle masses may be different, fermions have only one generation (See section 2.4.2.), and if our universe and the sister universe were generated as a vacuum fluctuation it is possible that the sister universe has anti-particle dominance just as our universe has particle dominance. See chapter 4 for a discussion of this possibility.

2.5.4 Sister Universe Gravitation

The theory of gravitation is the same in the sister universe as in our universe. However cosmological parameter values and mass distributions may be significantly different and there is no guarantee that it is a large and expanding universe. The origin of its Big Bang and its expansion (Dark Energy) may differ. Since our Big Bang theory uses a Planck distribution of $Y^{V}(x)$ quanta as Dark Energy and thus as the source of the expansion of our universe (See section 1.3.) it is possible a Black Body spectrum of $Y^{V}(y)$ quanta may be the Dark Energy of the sister universe with similar consequences.

Whether the energy-momentum tensor of the sister universe is that of a perfect fluid and whether it has a Robertson-Walker metric are open questions as well. We assume the sister universe has the same simple topology as our universe.

2.5.5 Sister Quantum Field Theory

We assume the quantum field theory of the sister universe is the same as that of our universe. Since quantum field theory gives similar results as the path integral formulation, and since we know of no alternative approach, and especially since QED calculations by T. Kinoshita (of Cornell University), his collaborators, and others have quantum theoretically confirmed experimental data with almost unbelievable accuracy, we think that the sister universe is undoubtedly also governed by quantum field theory. Thus we feel justified in our assumption.

3. Complex Coordinates Multiverses

Giordano Bruno suggested that earth was not alone and proposed that many worlds existed. In the early 20th century astronomers found nebulae were "island universes". We now know them to be galaxies. In the 19th century the possibility of extra dimensions and parallel universes was much discussed. William James[62] coined the word "multiverse" to describe parallel universes. This possibility was explored further theoretically in the latter part of the 20th century.[63]

The previous chapter, and other discussions, have strongly supported the possibility of a sister universe. In this chapter we begin discussing – not parallel universes – but a flat 16-dimensional space, the Flatverse, in which reside island universes. Our universe is one of these universes as is our sister universe. Some of the primary arguments in favor of the existence of a sister universe are:

1. It enables a mass scale to be determined through giving the Higgs lagrangian sector a mass term. The specification of a mass scale then determines distance and time scales as well as inertia and inertial reference frames.
2. It enables the definition of quantum gravity for our universe consistent with the Copenhagen interpretation of quantum theory which requires an independent observer for quantum processes. A sister universe being a macroscopic entity is an independent observer in principle.
3. It enables the formulation of a "clock" for quantum gravity – a necessity for a satisfactory theory of quantum gravity. The sister universe can play the role of a "clock."

---

[62] James, William, "Is Life Worth Living?", *Internat. Jrnl. Ethics* **6**, 10 (October, 1895).

[63] Max Tegmark, "Parallel Universes" arXiv:astro-ph/0302131 (2003); Zyga, Lisa, "Physicists Calculate Number of Parallel Universes", PhysOrg, 16 October 2009; Steven Weinberg, "Living in the Multiverse" (2005); George Ellis, "Does the Multiverse Really Exist?", *Scientific American* **305** (2): 38–43 (2011); and references therein.

The existence of a second universe, and the need for the Flatverse to properly define inertial reference frames, suggests that our universe exists within a multiverse that not only governs the phenomena of quantum gravity such as baby universes and wormholes within our universe but also governs phenomena in our sister universe and in the Flatverse.

This chapter describes quantum gravity based on two possible generalizations of Euclidean space to complex-valued coordinates and metrics, an analytic continuation generalization of general relativity to complex coordinates and metrics, and a non-analytic extension of general relativity to complex coordinates and metrics. In both cases we will use a sister universe as an observer and clock for our universe, the Reality group to transform complex coordinates and metrics to real-valued equivalents, and use the Flatverse as a platform for universes governed by quantum gravity.

Although the Flatverse is non-quantum[64] we shall introduce the possibility of an interaction that can propagate through the Flatverse between universes and also probe the contents of universes in chapter 4. We will be guided by an analogy with electromagnetic probes of protons: for low energies the probes treat the proton as an elementary particle with a spatial extension; for very high energies the probes "penetrate" the proton revealing its quark-parton constituents. Thus the quantum "observer" will use this interaction, in principle, if not in practice,[65] to observe quantum universe phenomena. Chapter 6 will propose a gauge boson interaction somewhat like QED but dependent on baryon number. Its existence would account for baryon number conservation (which is very accurately confirmed by experiment) and take advantage of the excess of baryons in our universe. Our universe has total charge zero and total energy zero. But it does have an excess of baryons. We will also suggest a possible reason for its excess numbers of baryons.

3.1 Physical Laws of the Multiverse Flatverse and its Universes

Since the multiverse Flatverse appears to have at least two universes within it and probably has many more the question arises: What are the physical laws of these universes. They may be the same. They may differ.

---

[64] But universes within the Flatverse are quantum.

[65] The strength of this interaction will be much, much weaker than gravity. See chapter 4.

Our primary assumption for the physical laws[66] of the multiverse is:

The physical laws of the multiverse (Flatverse) and the universes within it have the same form. The coupling constants and masses appearing in the physical laws of universes may be universe dependent.

The discussions and results presented from this point on are based on this assumption. The assumption is based on the generality of the origin of the form of our New Standard Model and of Quantum Gravity in complex space-time geometry as demonstrated in our earlier books and summarized in chapter 1.

The form of The New Standard Model and of Quantum Gravity was developed in a series of books which are largely summarized in the books listed in the Preface. Starting from a need for a theory supporting synchronized complex processes that leads to a four dimensional theory, and invariance under complex Lorentz transformations necessitated by the constancy of the speed of light (massless particles) in all inertial reference frames, we derive the form of the New Standard Model. Invariance under complex general relativistic transformations leads to Quantum Gravity and then to the Wheeler-DeWitt equation. Thus we believe the form of fundamental physical law is the same in all universes and in the Flatverse.

The masses and coupling constants of the New Standard Model and Quantum Gravity are not fixed by these geometric considerations. Therefore their values may differ in the various universes of the multiverse unless we can develop an understanding of their origin in geometry although another source for the origin of the values of masses and coupling constants is not excluded.

We view the universes as embedded in the Flatverse as described in chapters 1 and 2. The equation

$$z_i = f_i(x) \tag{1.4}$$

defines the extent of a universe in the Flatverse. The domain of 4-dimensional coordinates x of a universe specify a surface in the 16-dimensional Flatverse by

---

[66] Outside of universes gravitation is zero since every universe has a total energy of zero enabling Flatverse gravitation to be set to zero.

the range of Flatverse coordinate values z determined by eq. 1.4 assuming a completely classical theory. We have extended this theory to Quantum Gravity. Consequently the 16-dimensional surface of a universe is quantum (fuzzy) and determined by the quantum universe's wave function. This wave function is a solution of the Wheeler-Dewitt equation. The Wheeler-DeWitt equation assumes a single 4-dimensional universe with real-valued coordinates and metrics. Since our theory requires complex coordinates and thus complex-valued metrics, we must extend the Wheeler-DeWitt equation to complex metrics (next section). We must also extend it to 16-dimensional complex-valued coordinates and metrics. We perform this extension in chapter 4. The Wheeler-DeWitt equation then assumes a Flatverse form that we can apply uniformly to the universes within the Flatverse and to the 16-dimensional space between them.

As a result the universes are each described by a wave function of some complexity. The Flatverse is a flat space. Therefore the complete Flatverse wave function has the form

$$\psi_{\text{FlatverseTot}} = \prod_i \psi_i$$

where the product over i indicates the product of the wave functions of all the universes in the Flatverse. Loosely speaking we see the Flatverse as island universes floating in flat 16-dimensional space.

3.2 Analytically Continued Wheeler-DeWitt Equation to Complex Metrics under a Faddeev-Popov Method Restriction

In this section we extend the Wheeler-DeWitt equation[67] to complex coordinates and metrics by analytic continuation and impose the condition that metrics must be real-valued using the Faddeev-Popov Method. Our procedure will be to take the Wheeler-DeWitt equation for real-valued metrics (and coordinates), **analytically continue it to the case of complex metrics and coordinates,**[68] and then impose the condition that physically acceptable metrics must be real-valued via the Faddeev-Popov Method. Our motivation is two-fold:

---

[67] DeWitt, B. S., Phys. Rev. **160**, 1113 (1987).

[68] The piecewise analytic continuation of general relativity to complex coordinates and metrics is described in some detail in Blaha (2004). The gist of the continuation is that all equations have the same form after analytic continuation due to a basic theorem of complex mathematics that the analytic extension of equations from real values is unique.

we have shown that the form of The Standard Model of Particles can be derived from complex space-time considerations demonstrating that we exist in a "masked" complex space-time; and we have shown the imposition of a restriction on a gauge theory such as gravitation[69] should be implemented using the Faddeev-Popov Method or equivalent.

We start by noting the canonical decomposition of a real-valued metric $g_{\mu\nu}$ is defined by:

$$g_{\mu\nu}(x) = \eta_{\alpha\beta}\,\partial\omega^{\alpha}/\partial x^{\mu}\;\partial\omega^{\beta}/\partial x^{\nu} \qquad (3.1)$$

$$g_{\mu\nu} = g_{\nu\mu}$$

with inverse

$$g^{\mu\nu} = \eta^{\alpha\beta}\,\partial x^{\mu}/\partial\omega^{\alpha}\;\partial x^{\nu}/\partial\omega^{\beta} \qquad (3.2)$$

The decomposition of the real-valued metric is

$$g_{\mu\nu} = \begin{bmatrix} -\alpha^2\beta_k\beta^k & \beta_j \\ \beta_i & \gamma_{ij} \end{bmatrix} \qquad (3.3)$$

$$g^{\mu\nu} = \begin{bmatrix} -\alpha^{-2} & \alpha^{-2}\beta^j \\ \alpha^{-2}\beta^i & \gamma^{ij}-\alpha^2\beta^i\beta^j \end{bmatrix} \qquad (3.4)$$

where

$$\gamma_{ik}\,\gamma^{kj} = \delta_i^{\,j} \qquad\qquad \beta^i = \gamma^{ij}\,\beta_j \qquad (3.5)$$

The Wheeler-DeWitt equation <u>for real-valued metrics</u> is

$$(G_{ijkl}\,\delta/\delta\gamma_{ij}\,\delta/\delta\gamma_{kl} + \gamma^{½}\,{}^{(3)}R + 2\lambda\,\gamma^{½}\,{}^{(3)})\Psi({}^{(3)}\mathcal{G}) = 0 \qquad (3.6)$$

where λ is the cosmological constant, and where the Wheeler-DeWitt metric is

$$G_{ijkl} = ½\,\gamma^{-½}(\gamma_{ik}\gamma_{jl} + \gamma_{il}\gamma_{jk} - \gamma_{ij}\gamma_{kl}) \qquad (3.7)$$

---

[69] Such as real-valuedness.

The functional derivatives $\delta/\delta\gamma_{ij}$ have several interpretations that are presumably equivalent. DeWitt characterizes them as coordinate independent specifications of the 3-metric. The wave function $\Psi({}^{(3)}\mathcal{G}) = \Psi(\gamma_{ij})$, where ${}^{(3)}\mathcal{G}$ is a geometry, is *not* coordinate dependent. It is invariant under coordinate changes. ${}^{(3)}\mathcal{G}$ is a discrete infinity of independent invariants constructed from products of the Riemann tensor and its covariant derivatives.

Hartle and Hawking[70] derive the Wheeler-DeWitt equation from a path integral formalism for quantum gravity. Their path integral can be represented as

$$Z = N \int \delta g(x) \exp(iS_E[g]) \tag{3.8}$$

where S_E is the classical action for gravity and the functional integral is an integral over all 4-geometries. Changing to DeWitt's notation based on the spatial metric γ_{ij} and expressing eq. 3.8 in a more explicit form for the benefit of use in conjunction with the Faddeev-Popov Method we have

$$Z = N \int \sum_{i,j} \prod_{x} d\gamma_{ij}(x) \exp(iS_E[\gamma]) = N \int D\gamma \exp(iS_E[\gamma]) \tag{3.9}$$

The integrand, being a functional integral over all space, is independent of the coordinates.

The Wheeler-DeWitt equation applies to real-valued metrics $\gamma^{ij}(x)$. We now extend this equation to apply to complex-valued metrics using the local U(4) group. In doing this we realize that there are an infinite number of complex-valued metrics in the orbit corresponding to each real-valued metric.

This redundancy can be resolved by realizing the physical measurement of an invariant interval and the coordinates from which it is derived are always real-valued. Yardsticks and clocks can only measure real-valued numbers. Based on this physical principle we can generalize quantum gravity to complex coordinates and metrics by using the Faddeev-Popov method to constrain the set of paths in the quantum gravity path integral eq. 3.9. Using the Faddeev-Popov Method the constraint can be expressed in terms of an infinitesimal transformation of the metric to a complex value. Using an infinitesimal U(4) transformation V:

---

[70] Hartle, J. B. and Hawking, S. W., Phys. Rev. D **28**, 2960 (1983).

$$[\exp(ia_j(x'',x')U_j)]^\alpha{}_\mu = S(x'',x')^\alpha{}_\mu = \partial x''^\alpha/\partial x'^\mu \quad (3.10)$$

$$V = \exp(ia_k(x'',x')U_k) \cong I + ia_k(x'',x')U_k \quad (3.11)$$

$$V^\dagger = [\exp(ia_k(x'',x')U_k)]^\dagger \cong I - ia_k(x'',x')U_k \quad (3.12)$$

where $a_j(x'',x')$ is the jth real-valued local infinitesimal parameter, and U_k is one of the 16 hermitean generators of U(4). (We treat the index on a_k and U_k as lower case and sum on k.) We find the condition fixing the metric to a physical *real* value is:

$$F^{aij}(\gamma^a(x)) = \text{Im}\, \gamma^{ija}(x) \equiv -½\, i(\gamma^{ija}(x) + \gamma^{ija}(x)^*) = 0 \quad (3.13)$$

where

$$F^{aij}(\gamma^a(x)) = \text{Im}\{(\delta^i{}_m + ia_k(x',x)U_k{}^i{}_m)\gamma^{mp}(x)(\delta^j{}_p + ia_k(x',x)U_k{}^j{}_p)\} \quad (3.14)$$

based on the infinitesimal form

The Reality condition eq. 3.13 is implemented within the path integral formalism with the Faddeev-Popov Method identity

$$1 = \int D\gamma\, \Delta(F(\gamma))\, \delta(\delta F^{aij}(\gamma^a(x))/\delta a_n(x',x)) = \int D\gamma\, \Delta(F(\gamma))\, \delta(F(\gamma^{ij})) \quad (3.15)$$

Eq. 3.14 yields

$$\begin{aligned}\delta F^{aij}(\gamma^a(x))/\delta a_n(x',x)\,|_{a=0} &= \text{Im}\{\delta_{kn} i U_k{}^i{}_m \gamma^{mp}(x)\delta^j{}_p + \delta_{kn}\delta^i{}_m i\gamma^{mp}(x)U_k{}^j{}_p]\} \\ &= \text{Re}\{\gamma^{mp}[[U_n{}^i{}_m\delta^j{}_p + \delta^i{}_p U_n{}^j{}_m]\} \\ &= \gamma^{mp}(x)\text{Re}\{U_n{}^i{}_m\delta^j{}_p + \delta^i{}_p U_n{}^j{}_m\} = \gamma^{mp}(x)\xi^{ij}{}_{nmp}\end{aligned} \quad (3.16)$$

since $\gamma^{mp}(x)$ is made real by the $\delta(\text{Im}\,\gamma^{mp}(x))$ where

$$\xi^{ij}{}_{nmp} = \text{Re}\{U_n{}^i{}_m\delta^j{}_p + \delta^i{}_p U_n{}^j{}_m\} \quad (3.17)$$

Note $\xi^{ij}{}_{nmp}$ is symmetric in i and j, and becomes effectively symmetric in m and p when combined with $\gamma^{mp}(x)$ in eq. 3.16. Calculating $\Delta(F(\gamma))$ we obtain

$$\Delta(F(\gamma)) = [\det \delta F^{aij}(\gamma^a(x))/\delta a_n(x',x)\,|_{a=0}]^{-1} \quad (3.18)$$

We can rewrite this Faddeev-Popov determinant as a path integral over an anti-commuting c-number scalar field χ with a ghost Lagrangian:

$$\Delta(F(\gamma)) = \int D\chi^* D\chi \exp[i\int d^4x\, \gamma^{1/2} \mathscr{L}_\gamma{}^{ghost}(x)] \qquad (3.19)$$

where

$$\mathscr{L}_\gamma{}^{ghost}(x) = \chi^*(x)[U_{nij}\xi^{ij}{}_{nmp}\gamma^{mp}(x)]\chi(x)$$

$$= \chi^*(x)(\text{Re } U_{nij})\xi^{ij}{}_{nmp}\gamma^{mp}(x)]\chi(x) \qquad (3.20)$$

$$= \chi^*(x)\chi(x)\xi_{mp}\gamma^{mp}(x) \qquad (3.21)$$

since $\xi_{mp}\gamma^{mp}(x)$ is a c-number:

$$\xi_{mp} = (\text{Re } U_{nij})\xi^{ij}{}_{nmp} \qquad (3.22)$$

Using the hermitean 4 × 4 U(4) generators we find ξ_{mp} Is a diagonal matrix and has the value

$$\xi_{mp} = \xi_m\delta_{mp} \qquad (3.22a)$$

where the diagonal elements are

$$\xi_0 = 8$$
$$\xi_1 = 4$$
$$\xi_2 = 2$$
$$\xi_3 = 0 \qquad (3.22b)$$

and the non-diagonal elements are zero.

The Faddeev-Popov generated terms when added to the Einstein Action appear to have important ramifications – particularly with respect to the Cosmological Constant. We explore that issue next.

It also appears to generate directionality in space that seems to be similar to the unusual distribution of matter seen in the universe. We discuss that topic next.

3.3 Possible Source of the Cosmological Constant in the Complex Space-time – Faddeev-Popov Constraint Term

The Wheeler-DeWitt equation

$$(G_{ijkl}\,\delta/\delta\gamma_{ij}\,\delta/\delta\gamma_{kl} + \gamma^{\frac{1}{2}\,(3)}R + 2\lambda\,\gamma^{\frac{1}{2}\,(3)})\Psi({}^{(3)}\mathcal{G}) = 0 \qquad (3.23)$$

has the cosmological constant, Λ, as one of its terms. This equation is derived from the Hamiltonian constraint that ultimately follows from the Einstein action. Inserting the Faddeev-Popov term in the Einstein lagrangian yields

$$S_E(\gamma) = -(16\pi G)^{-1}\int d^4x\,\gamma^{\frac{1}{2}}\{R(x) - 2\lambda + \mathcal{X}^*(x)\mathcal{X}(x)\xi_{mp}\gamma^{mp}(x)\} \qquad (3.24)$$

The constant matrix ξ_{mp} is a product of parts of the generators of U(4) given by eq. 3.22a:

$$\xi_{mp} = (\mathrm{Re}\,U_{nip})\,(\mathrm{Re}\,U_{n\,m}^{\;i}) \qquad (3.25)$$

We will now show that the term

$$\Pi = \gamma^{\frac{1}{2}}\mathcal{X}^*(x)\mathcal{X}(x)\xi_{mp}\gamma^{mp}(x) \qquad (3.26)$$

upon variation of the metric $\delta\gamma_{\mu\nu}$ gives a term which is approximately a cosmological constant term assuming $\mathcal{X}(x)$ is approximately constant (with its implied divergence eliminated by renormalization of the path integral), and assuming an almost flat space-time $\gamma_{\mu\nu} \cong \eta_{\mu\nu}$. Varying the metric for Π yields

$$\delta\Pi = \delta\gamma_{\mu\nu}\,\{\tfrac{1}{2}\,\gamma^{\frac{1}{2}}\mathcal{X}^*(x)\mathcal{X}(x)\xi_{mp}\gamma^{mp}(x)\gamma^{\mu\nu} - \gamma^{\frac{1}{2}}\mathcal{X}^*(x)\mathcal{X}(x)\xi_{mp}\gamma^{m\mu}\gamma^{p\nu}\} \qquad (3.27)$$

The resulting modified Einstein field equation is

$$R^{\mu\nu} - \tfrac{1}{2}\,g^{\mu\nu}R + \lambda\,\gamma^{\mu\nu} + \tfrac{1}{2}[\gamma^{mp}(x)\gamma^{\mu\nu} - \gamma^{m\mu}\gamma^{p\nu}]\xi_{mp}\mathcal{X}^*(x)\mathcal{X}(x) = -8\pi T^{\mu\nu} \qquad (3.28)$$

The term $\tfrac{1}{2}\gamma^{mp}(x)\xi_{mp}\mathcal{X}^*(x)\mathcal{X}(x)\gamma^{\mu\nu}$ appearing above is, or is a contribution to, the cosmological constant assuming a nearly flat universe as our universe seems to be. Thus $\gamma_{\mu\nu} \cong \eta_{\mu\nu}$ to good approximation and $\mathcal{X}^*(x)\mathcal{X}(x)$ can be taken to be constant since the time derivative of $\mathcal{X}(x)$ does not appear in the lagrangian. Consequently

$$\lambda_{tot}^{\;\mu\nu} \cong \lambda g^{\mu\nu} + \mathcal{X}^*(x)\mathcal{X}(x)g^{\mu\nu} = (\lambda + \lambda_{F\text{-}P})\,g^{\mu\nu} \qquad (3.29)$$

by eq. 3.22b.

Given the somewhat problematic state of our understanding of the cosmological constant it is not impossible that the complexity of space-time leading to $\lambda_{F\text{-}P}$ may be the sole origin of the cosmological constant. In evaluating eq. 3.29 we may normalize $\chi^*(x)\chi(x) = 1$ by adjusting the overall normalization of the path integral since $\chi(x)$ is time independent. Then if the "bare" cosmological constant is zero we obtain

$$\lambda_{tot} = \lambda_{F\text{-}P} = 1 \tag{3.30}$$

by eq. 3.29.

If this is true then we have achieved a space-time origin for the cosmological constant rather than an ad hoc origin. The modified Einstein field equation then is

$$R^{\mu\nu} - ½\, g^{\mu\nu}R + \lambda_{tot}\gamma^{\mu\nu} - ½\, \gamma^{m\mu}\gamma^{p\nu}\xi_{mp} = -8\pi T^{\mu\nu} \tag{3.31}$$

by eqs. 3.28 and 3.30. The 4th term in eq. 3.31 introduces an additional feature of the evolution of the universe: the expansion is not the same in all directions. It is different in different space-time directions. If we look at the diagonal components of eq. 3.31 we find:

$$R^{00} - ½\, g^{00}R + \lambda_{tot} - 4 = R^{00} - ½\, g^{00}R - 3 = -8\pi T^{00} \tag{3.32}$$

$$R^{11} - ½\, g^{11}R + \lambda_{tot} - 2 = R^{11} - ½\, g^{11}R - 1 = -8\pi T^{11} \tag{3.33}$$

$$R^{22} - ½\, g^{22}R + \lambda_{tot} - 1 = R^{22} - ½\, g^{22}R = -8\pi T^{22} \tag{3.34}$$

$$R^{33} - ½\, g^{33}R + \lambda_{tot} - 0 = R^{33} - ½\, g^{33}R + 1 = -8\pi T^{33} \tag{3.35}$$

based on eq. 3.22b, approximating $\gamma_{\mu\nu} \cong \eta_{\mu\nu}$ only in the 3rd and 4th terms of eqs. 3.28 and 3.31. We thus find differing cosmological constants ("expansion" rates) in different directions.

3.4 Deviations from Uniform Expansion of the Universe Due to a Directional Cosmological Constant

Deviations from uniform evolution in the universe have been noted by several astrophysical experiments. Recently in September, 2013 data from

NASA's Wilkinson Microwave Anisotropy Probe (WMAP) has been analyzed and suggests a "saddle-shaped" universe (in space and time) based on a major anomaly in the Big Bang's afterglow. A recent study of Liddle and Cortês[71] suggests that the universe while almost flat may be marginally open. *The above differences in cosmological constant values for different directions gives a straight-forward way of understanding the origin of a generally saddle-shaped universe.*

3.5 Impact of the Faddeev-Popov Complexity Term on the Wheeler-DeWitt Equation

The Faddeev-Popov term that arises because of the restriction of the metrics and coordinates to real values also impacts on the Wheeler-DeWitt equation since it is derived from the lagrangian via the Hamiltonian it generates. The Wheeler-DeWitt equation changes to

$$(G_{ijkl}\ \delta/\delta\gamma_{ij}\ \delta/\delta\gamma_{kl} + \gamma^{\frac{1}{2}\,(3)} R + \{2\lambda + 3\gamma^{mp}\xi_{mp}'\}\ \gamma^{\frac{1}{2}\,(3)})\Psi({}^{(3)}\mathcal{G}) = 0 \qquad (3.36)$$

where $\xi_{mp}' = \text{diag}(1, 0, -1)$ is from the spatial 3×3 part of ξ_{mp}. Eq. 3.36 becomes

$$(G_{ijkl}\ \delta/\delta\gamma_{ij}\ \delta/\delta\gamma_{kl} + \gamma^{\frac{1}{2}\,(3)}R + \{2\lambda + 12\gamma^{11} + 6\gamma^{22}\}\gamma^{\frac{1}{2}\,(3)})\Psi({}^{(3)}\mathcal{G},\ \mathcal{L}_F) = 0 \qquad (3.37)$$

where λ, which appears in eq. 3.29, may be zero. λ_{tot} is the effective cosmological constant. We have introduced the "lagrangian" $\mathcal{L}_F$ representing the surroundings of the universe anticipating that a universe is a surface contained within a larger structure, the Flatverse. Thus $\mathcal{L}_F$ is a specification of the surrounding Flatverse environment.

The additional Faddeev-Popov terms in eq. 3.37 change the solutions for the wave function $\Psi({}^{(3)}G, L_F)$ to that of asymmetric metrics due to the linear terms in γ^{11} and γ^{22}.

The functional Wheeler-DeWitt equation of eq. 3.37 resembles a Klein-Gordon equation.[72] Solutions of this equation can be expressed as path integrals:

---

[71] Liddle, A. R. and Cortês, M., Phys. Rev. Lett. **111**, 111302 (2013).

[72] Hartle, J. B. and Hawking, S. W., Phys. Rev. D **28**, 2960 (1983).

$$\Psi({}^{(3)}\mathcal{G}, \mathcal{L}_F) = N \int_{\mathcal{C}} \delta g(x) \exp(-I(g, \mathcal{L}_F)) \tag{3.38}$$

where $I(g, \mathcal{L}_F)$ is the effective total Euclidean action for the open universe case. See Hartle and Hawking for a detailed study in the case of a symmetric cosmological constant.

From eqs. 3.33 – 3.35 it appears that there could be expansion in the 3-direction, contraction in the 1-dimension and an ambivalent situation in the 2-direction.

It does not appear that the issues of the Wheeler-DeWitt equation are resolved by the extended Extended Wheeler-DeWitt equation presented here:

- Divergences in integrals in inner products, thus requiring renormalization.
- Negative probabilities in inner products,
- Issues with the requirement of space-like surfaces,
- The frontier divergence singularity.

Later we consider another form of the Wheeler-DeWitt equation expressed in terms of Flatverse geometry. We will reconsider the issues of the above formulation again in the Flatverse.

3.5.1 The Universe's Web of Galaxies and the New Faddeev-Popov Terms in the Wheeler-DeWitt Equation

The $12\gamma^{11} + 6\gamma^{22}$ terms in eq. 3.37 introduce a spatial directionality that might be related to the web-like structure of galaxies called the cosmic web that has been found in our universe. The most recent results on this *universal web* which connects all galaxies in the universe are presented by Cantalupo et al in Nature (Jan. 19, 2014). Their results stand in strong contrast to the original expectation of randomly distributed galaxy clusters.

The additional terms in eq. 3.37 strongly suggest broken linear structures that could become web-like as the universe evolved since its beginning.

3.6 Is there a Spatial Asymmetry in our Universe Due to the Metric Restriction to Real Values?

The considerations of subsections 3.2 and 3.4 suggest a spatial asymmetry in the universe. Eq. 3.33 suggests possible contraction (or slow expansion) in the 1-direction.

$$R^{11} - \tfrac{1}{2} g^{11}R + \lambda_{tot} - 2 = R^{11} - \tfrac{1}{2} g^{11}R - 1 = -8\pi T^{11} \qquad (3.33)$$

An analysis[73] of The Sloan Digital Sky Survey data shows the existence of an extremely long, large quasar group powered by ultra-massive black holes that extends 4 billion light-years and is 1.6 billion light-years in breadth in other directions. This is the largest known structure in the universe. It might reflect the asymmetry in the 1-direction with expansion in the 2-direction and 3-direction but slow expansion in the 1-direction leading to a line of closely spaced quasars in the 1-direction.

Another indication of asymmetry in the universe is a large dense spot in the universe found by the recent European Planck satellite experiment. This region of high density is difficult to understand in conventional theories of our expanding universe.

[73] Roger Clowes et al, Monthly Notices of the Royal Astronomical Society, Jan. 11, 2013.

4. Island Universes in the Flatverse - Multiverse Physiognomy

In chapters 1 - 3 we described the embedding of our curved universe within the 16-dimensional Flatverse. The embedding equations were

$$z_i = f_i(x) \tag{1.4}$$
$$ds^2 = g_{ij}dz^i dz^j \tag{1.5}$$
$$g_{ij} = g_{ji} \tag{1.6}$$

for the Flatverse. The metric tensor $g_{\mu\nu}$ of the universe is

$$g_{\mu\nu} = \partial f_j/\partial x^\mu \, \partial f_j/\partial x^\nu \tag{1.7}$$

with an implied sum over the subscript j.[74] Outside of closed universes, which we assume are the only non-zero aggregations of mass-energy, the Flatverse is truly flat with a Euclidean metric $g_{ij} = 1$ for all i = j and zero otherwise.

This type of embedding has been known since the early 20th century. The embeddings then were usually within a 10-dimensional flat space since the 4-dimensional real-valued metric has 10 independent components. We require a 16-dimensional Flatverse for a complex $g_{\mu\nu}$ due to its 16 independent components (32 elements forming a symmetric array).

However this embedding has unstated implications when there is a set of universes within a Flatverse multiverse.

4.1 Implications of a Multiverse within the Flatverse - Island Universes

The implication of an embedding of the type of eq. 1.4 in the Flatverse is its apparently static nature. A universe occupies a region within the Flatverse. Its

[74] Recall that the complex four-dimensional metric has 16 independent real-valued components thus leading to an 8-dimensional complex embedding within the 16-dimensional complex Flatverse.

gravitational field and mass-energy are confined to the universe's region – rather like quarks and gluons are confined within hadrons.[75] We will consider the confinement of gravitation in more detail later.

From the Flatverse coordinates viewpoint a universe can change its location and/or size. Universes can expand or contract. Within a universe, its size and mass-energy distributions may change.

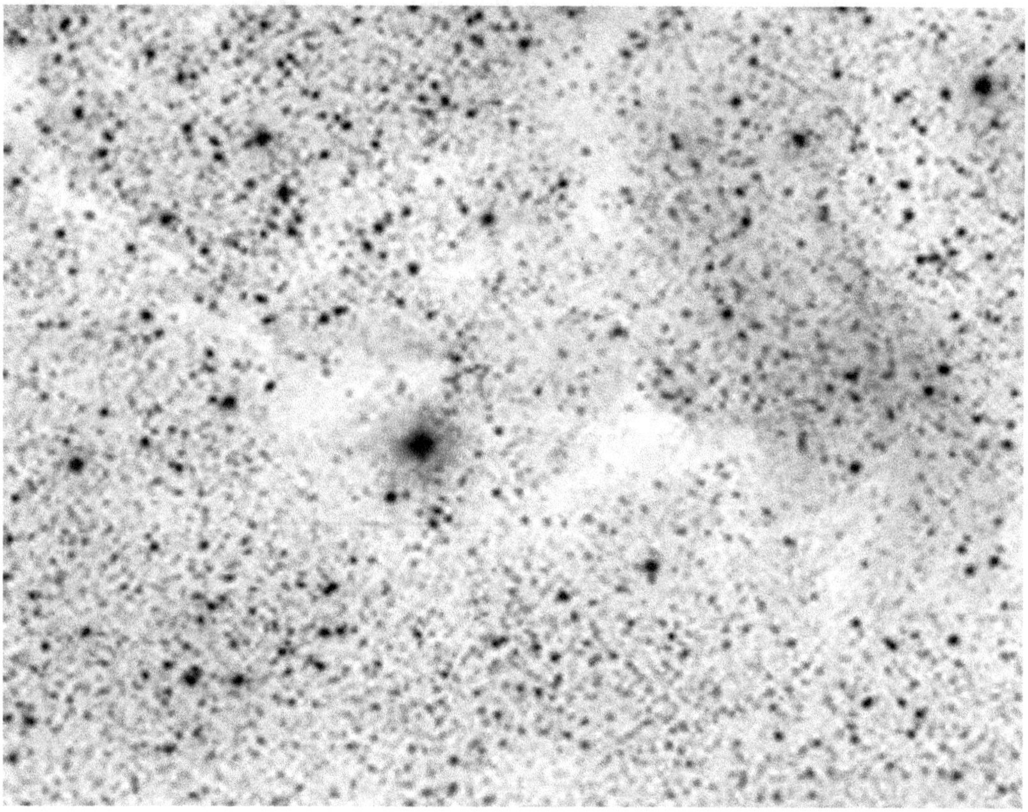

Figure 4.1. A symbolic multiverse visualization of universes (black spots) within the Flatverse.

---

[75] The possibility that the gravitational field and the strong interaction gauge field might have similarities such as confinement and a linear potential was first considered by this author in "Quantum Gravity and Quark Confinement" by Stephen Blaha, Lett. Nuovo Cim. 18:60, 1977. Honorable Mention in the Gravity Research Foundation Essay Competition in 1978.

We thus arrive at a view of the Flatverse multiverse as a set of island universes scattered throughout the Flatverse. See Fig. 4.1 for a visualization.

This view is in accord with the observation that the total mass-energy contained within a universe is zero (the mass-energy of matter and radiation being cancelled by the universe's gravitational energy), and the total 3-momentum contained within a universe is also zero.

However within the Flatverse a universe can have a non-zero energy and momentum as an entity. This energy-momentum does not arise from the contents of the universe. For example a universe can move within the Flatverse as a whole with all parts of the universe uniformly participating in this motion. We consider this possibility in more detail later.

4.2 Transformation of the Wheeler-Dewitt Equation to the Flatverse

The Wheeler-DeWitt equation for a universe is in accord with the views expressed in section 4.1. This equation is expressed in terms of a universe's metric and applies only within the universe. It is possible to transform it to a Flatverse form using eq. 1.7 and extend it to the entire Flatverse. Then universes become islands of Wheeler-DeWitt dynamics surrounded by flat space in which the Wheeler-DeWitt dynamics is trivial.

The extended Wheeler-DeWitt equation including Faddeev-Popov terms is

$$(G_{ijkl}\, \delta/\delta\gamma_{ij}\, \delta/\delta\gamma_{kl} + \gamma^{½}\, {}^{(3)}R + \{2\lambda + 12\gamma^{11} + 6\gamma^{22}\}\gamma^{½}\, {}^{(3)})\Psi({}^{(3)}\mathcal{G}, \mathcal{L}_F) = 0 \quad (3.37)$$

where $G_{ijkl} = ½\, \gamma^{-½}(\gamma_{ik}\gamma_{jl} + \gamma_{il}\gamma_{jk} - \gamma_{ij}\gamma_{kl})$. We will outline the salient features of the transformation to a Flatverse equivalent using eq. 1.7, the interpretation of the functional integral symbolic form (see eq. 3.9 for an example), and the corresponding form that functional differentiation must have. The functional path integral symbolic form is a shorthand notation for

$$\int \sum_{i,j} \prod_{x} d\gamma_{ij}(x) \equiv \int D\gamma \qquad (4.1)$$

Similarly the functional derivatives in eq 3.37 are also a shorthand notation:

$$\delta/\delta\gamma_{ij} \equiv \prod_{x} \sum_{k} \partial/(\partial f_k/\partial x^i\, \partial f_k/\partial x^j) \qquad (4.2)$$

using eq. 1.7. The derivative in eq. 4.2 can be re-expressed using the chain rule of differentiation as

$$\partial/\partial(\partial f_k/\partial x^i\,\partial f_k/\partial x^j) = [\delta^m{}_j\partial f_k/\partial x^i + \delta^m{}_i\partial f_k/\partial x^j\,]^{-1}\,\partial/\partial(\partial f_k/\partial x^m) \qquad (4.3a)$$

and

$$\partial/\partial(\partial f_k/\partial x^m) = (\partial^2 f_k/\partial x^{m2})^{-1}\partial/\partial x^m \qquad (4.3b)$$

with no implied summations over k or m.

Thus the functional notation $\delta/\delta\gamma_{ij}$ becomes, using eqs. 4.2 and 4.3,

$$\delta/\delta\gamma_{ij} \equiv \prod_x\sum_k \partial/(\partial f_k/\partial x^i\,\partial f_k/\partial x^j) \equiv \prod_x\sum_k [\delta^m{}_j\,\partial f_k/\partial x^i + \delta^m{}_i\,\partial f_k/\partial x^j\,]^{-1}\partial/\partial(\partial f_k/\partial x^m) \qquad (4.4a)$$

$$\equiv \prod_x\sum_k \{[\delta^m{}_j\,\partial f_k/\partial x^i + \delta^m{}_i\,\partial f_k/\partial x^j\,](\partial^2 f_k/\partial x^{m2})\}^{-1}\,\partial/\partial x^m \qquad (4.4b)$$

with summations over k and m = 1, 2, 3.

The substitution of this form for the Wheeler-DeWitt functional derivatives, and the use of eq. 1.7 for all metric terms in eq. 3.37 above, yields a Flatverse equivalent to the Wheeler-DeWitt equation for universe metrics.

4.3 Flatverse Wheeler-Dewitt Equation

The Flatverse Wheeler-DeWitt equation for a universe is

$$\Big(G_{ijkl}\Big\{\prod_x\sum_m\{[\delta^m{}_j\,\partial f_n/\partial x^i + \delta^m{}_i\,\partial f_m/\partial x^j\,](\partial^2 f_n/\partial x^{m2})\}^{-1}\partial/\partial x^m\Big\}\Big\{\prod_x\sum_m\{[\delta^m{}_k\,\partial f_n/\partial x^l + \delta^m{}_l\,\partial f_n/\partial x^k\,](\partial^2 f_n/\partial x^{m2})\}^{-1}\partial/\partial x^m\Big\} + \gamma^{1/2\,(3)}R + \{2\lambda + 12\gamma^{11} + 6\gamma^{22}\}\gamma^{1/2\,(3)}\Big)\Psi({}^{(3)}\mathcal{G}, \mathcal{L}_F) = 0 \qquad (4.5a)$$

or in terms of Flatverse coordinates y_n

$$0=\Big(G_{ijkl}\Big\{\prod_x\sum_m\{[\delta^m{}_j\,\partial y_n/\partial x^i + \delta^m{}_i\,\partial y_m/\partial x^j\,](\partial^2 y_n/\partial x^{m2})\}^{-1}\partial/\partial x^m\Big\}\Big\{\prod_x\sum_m\{[\delta^m{}_k\,\partial y_n/\partial x^l + \delta^m{}_l\,\partial y_n/\partial x^k\,](\partial^2 y_n/\partial x^{m2})\}^{-1}\partial/\partial x^m\Big\} + \gamma^{1/2\,(3)}R + \{2\lambda + 12\gamma^{11} + 6\gamma^{22}\}\gamma^{1/2\,(3)}\Big)\Psi({}^{(3)}\mathcal{G}, \mathcal{L}_F) \qquad (4.5b)$$

where the sums over n and m in each pair of {} are done independently. All references to the metric are expressed in terms of the Flatverse using eq. 1.7.

The Flatverse outside of universes is trivial due to the flatness of the Flatverse.

Due to the products over all coordinates, the Flatverse expression for $\delta/\delta\gamma_{ij}$ is independent of x (and Flatverse coordinates) in accordance with the space-time independence of the original Wheeler-DeWitt equation.

The Flatverse form of the Wheeler-DeWitt equation directly relates the metric of a universe to the Flatverse. Every universe has two sets of coordinates: coordinates, usually labeled x, embodying the curvature of the universe induced by gravitation, and Flatverse coordinates usually labeled y. Fig. 4.2 symbolically depicts part of the Flatverse.

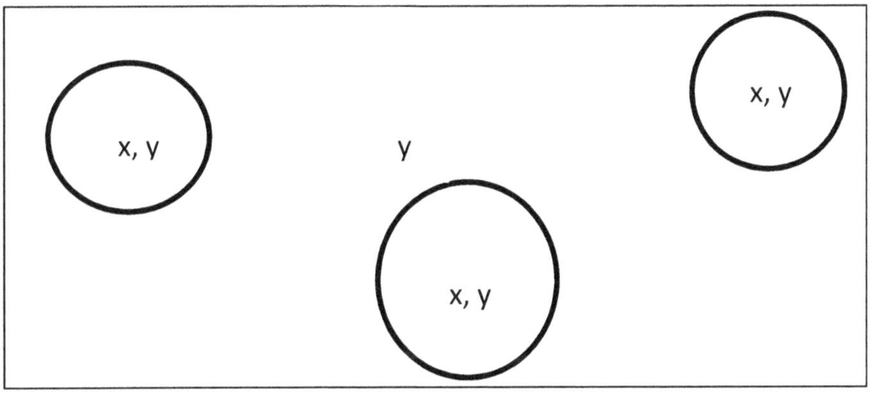

Figure 4.2. A symbolic view of part of the Flatverse with universes depicted as circles. Each universe has its own curvilinear coordinate 4-vector denoted x. The Flatverse coordinates are a 16-vector y. A Wheeler-DeWitt equation applies within each universe. The Wheeler-DeWitt equation for the Flatverse is trivial.

4.4 Flatverse Metric Functional Integral

Corresponding to the functional derivative in eq. 4.4 is an explicit Flatverse form of the functional integral

$$\int D\gamma \equiv \int \prod_x \sum_{i,j} d\gamma_{ij}(x) = \int \prod_x \sum_{i,j,k} d(\partial f_k/\partial x^i \, \partial f_k/\partial x^j) \tag{4.6}$$

$$= \int \prod_x \sum_{i,j,k} [d(\partial f_k/\partial x^i)\, \partial f_k/\partial x^j + d(\partial f_k/\partial x^j)\, \partial f_k/\partial x^i]$$

$$= 2\int \prod_x \sum_{i,j,k} d(\partial f_k/\partial x^i)\, \partial f_k/\partial x^j$$

$$= 2\int \prod_x \sum_{i,j,k} dx(\partial^2 f_k/\partial x^{i2})\, \partial f_k/\partial x^j \equiv 2 \int \delta f(x) \qquad (4.7a)$$

$$= 2\int \prod_x \sum_{i,j,k} dx(\partial^2 y_k/\partial x^{i2})\, \partial y_k/\partial x^j \equiv 2 \int \delta f(x) \qquad (4.7b)$$

The third line is due to the summation over i and j. The factor of 2 can be absorbed in the normalization factor. Eqs. 4.6 and 4.7, like the Wheeler-DeWitt equation, are independent of the coordinates due to the product over coordinates x. Thus the solution of the Wheeler-DeWitt equation takes a form similar to eq. 3.38 upon absorption of the factor of 2 into the normalization constant N

$$\Psi({}^{(3)}\mathcal{G}, \mathcal{L}_F) = N \int \delta f(x) \exp(-I(g, \mathcal{L}_F)) \qquad (4.8)$$

The real-valued metric, Flatverse version of the Wheeler-DeWitt equation differs from its conventional version by depending on $\partial f_k/\partial x^i = \partial y_k/\partial x^i$ which is a 16 by 3 = 48 component construct. The conventional version depends on the metric γ_{ij} which is a 3 by 3 = 9 component construct with 6 independent components due to its symmetry. The 48 components of $\partial y_k/\partial x^i$ are constrained to a 3-dimensional surface by 13 equations specifying the surface of the universe

$$H_a(y_1, y_2, \ldots, y_{16}) = 0 \qquad (4.9)$$

for a = 1, 2, …, 13. As a result the 48 components of $\partial y_k/\partial x^i$ have 9 independent components that are further constrained by the ij symmetry of eq. 4.5 to 6 independent components just as the metric γ_{ij}.

The complex-valued metric versions of the Wheeler-DeWitt equation have 9 independent metric components.

4.5 Flatverse Wheeler-DeWitt Solutions

The Flatverse Wheeler-DeWitt equation eq. 4.5b is for the case of real-valued metrics and coordinates. It clearly is asymmetric due to the $(12\gamma^{11}$ +

$6\gamma^{22}\}\gamma^{½}\ ^{(3)}$ term resulting from the application of the Faddeev-Popov method. As we discussed in section 3.6 there are observed asymmetries in the universe that may result from our extended Wheeler-DeWitt equation. While one could look for approximate solutions such as DeWitt, and Hartle and Hawking, have done it does not appear that much can be learned from those approaches in the present case. Therefore we will consider general aspects of universes that are implied by these equations.

4.5.1 Tachyonic Solutions of Wheeler-DeWitt Equation

The original Wheeler-DeWitt equation and its Flatverse equivalent (eq. 4.5) have a form that is similar in many respects to the Klein-Gordon equation. In particular as DeWitt[76] noted "a Klein-Gordon equation with $-\gamma^{½}\ {}^{(3)}R$ (our notation) playing the role of a mass-squared term. An important difference, however, is that ${}^{(3)}R$ can be either positive or negative, and hence the wave propagation of the state functional is not confined to time-like directions." DeWitt proceeds later in his paper to exclude consideration of negative "mass" squared terms.

In our view the negative "mass" squared cases represent tachyonic solutions of the Wheeler-DeWitt equation and should be considered as having the same validity as the positive mass squared solutions.

G_{ijkl} can be regarded as the contravariant metric of a 6-dimensional Riemannian manifold M with hyperbolic signature (-1, 1, 1, 1, 1, 1) with a time-like coordinate.[77] Tachyonic solutions are part of the set of solutions of the Wheeler-DeWitt equation. In the Flatverse formulation the tachyonic solutions are embedded as "islands" in the emptiness of the Flatverse.

If we use the Flatverse form of the Wheeler-DeWitt equation eq. 4.5b then the changes in time (dilations of γ_{ij} or equivalently dilations of y_n) can be viewed as specifying the overall motion of entire universes. We see now that we can have "normal" or tachyonic motion of universes. Clearly we are building towards a particle view of universes.[78]

---

[76] DeWitt, p. 1124.

[77] In the case of complex-valued metrics the Riemannian manifold would be 9-dimensional with signature (-1, 1,1,1,1,1,1,1,1).

[78] The Reality groups of 4-dimensional universes, and of the Flatverse, play a role in the physical interpretation of Flatverse phenomena.

4.5.2 Problems in the Solutions of the Wheeler-DeWitt Equation

There are a number of problem areas associated with the original and our extended Wheeler-DeWitt equations. DeWitt identified most of them in his paper, referenced earlier. While one could view these apparent problems as negatives, we will take the view that they are indicators of a deeper structure of universes below the Wheeler-DeWitt solutions just as the Dirac equation appeared to resolve analogous difficulties with the Klein-Gordon equation.

4.5.2.1 Negative Frequencies and Probabilities – Anti-Universes

DeWitt noted the existence of negative frequencies and negative probabilities associated with the solutions of the Wheeler-DeWitt equation. The rather close analogy to the Klein-Gordon equation in whose solutions similar issues appear is suggestive.

It appears that two general types of universes are embodied in the Wheeler-DeWitt equation. One type, which we call "normal", consists of universes like ours which have an excess of baryons (and consequently electrons to make an overall charge neutral universe). The other type of universe we will call *anti-universes*. These universes have an excess of anti-baryons (and positrons). Earlier we have suggested that our universe is paired with an anti-universe via a wormhole. We hypothesize the pair results from a vacuum fluctuation in the Flatverse described in some detail in chapter 6.

In direct analogy with the Klein-Gordon equation the negative frequency and negative probability issues are thereby resolved.

One can speculate that the Wheeler-Dewitt equation, which is effectively second order in the "time" derivative, can be factored into first derivative equations – perhaps through the introduction of more degrees of freedom in a fashion similar to Dirac's introduction of spinors – to achieve first order equations in "time." Then we would have to face the issue of interpreting "Dirac-like" Wheeler-DeWitt equations.[79] We could again fall back on the rationale for introducing spinors for Standard Model fermions – Asynchronous Logic – and suggest that universes embody a 4-valued Logic – although the concept of universes as logic values is somewhat strange. It appears premature to envision universes with logic "spin."

---

[79] We will not consider factoring the Wheeler-DeWitt equation in this volume.

Therefore we will be content, for the present, to assume both universes and anti-universes exist. An anti-universe will be presumed to be a universe in the Flatverse where anti-particles predominate. We will postulate that universes, and anti-universes like protons and anti-protons, are always bodies with extension and not point-like. Blaha (2013) provides an example of a Big Bang where a universe begins with extension and is not point-like.

Chapters 5, 6 and 7 describe a proposed dynamics of universes and anti-universes based on an abelian Flatverse gauge field.

4.6 General Types of Universes within the Flatverse

We determined the dimensionality of our universe based on principles of Asynchronous Logic that suggested a 4-valued logic that could be embodied in a 4-dimensional spinor matrix formulation. This 4-dimensional spinor formulation led to a 4-dimensional space-time. The requirement that the speed of light is the same in all inertial reference frames, and that transformations between reference frames in faster than light relative motion are physical, led to the requirement of complex coordinates and the complex Lorentz transformation group. The reality of all physical time and distance measurements led to the introduction of the Reality group that mapped complex quantities to real physical values. This chain of logic is in accord with Leibniz's abstract minimax principle that nature uses the simplest means to create complex physical phenomena.

While the preceding paragraph applies very nicely to our universe and our sister universe (introduced in Blaha (2013) the question of other possible types of universes within the multiverse naturally arises. Stars and galaxies have many varieties. Why should all universes be the same?

Having developed the fundamental nature of our universe from Logic (the only sure requirement of any physical theory) it seems reasonable to classify possible universes based on their fundamental logic. In Blaha (2011c) we developed matrix formulations for many-valued logics. Assuming no separate clock mechanism to synchronize parts of complex processes, we developed an n×n matrix formalism for n-valued logic.

Therefore we can develop a principal sequence of types of universes based on n-valued logic. We can summarize the small n-valued cases as follows:

n-Valued Logic	Matrix Representation Size	Spinor Components	Space-Time Dimensionality[80]
1	1×1	1	1
2	2×2	2	2
3	3×3	2	3
4	4×4	4	4
5	5×5	4	5
6	6×6	8	6

Table 4.1. Space-time dimensionality and number of spinor components corresponding to various n-valued logics.

Fant (2005) points out that VLSI circuits with spatially separated parts, which require time synchronization of activity without clocks, need a 4-valued logic at minimum. Thus for a complex universe such as ours the minimum space-time dimensionality is 4. For a smaller number of dimensions the complexity of physical processes is much diminished as the many solvable models of low space-time dimensionality show. Easily solved – not very complex phenomena!

Smaller dimensioned universes may well exist – but not with the richness of complexity that leads to our type of universe's phenomena such as life.

Larger dimension universes may well also exist. They would have an excess of phenomena that might preclude life as we know it.

The general tendency of physical phenomena to be largely based on extrema suggests that 4-dimensional space-time based on Leibniz's minimax principle is the "logical" choice.

The above classification scheme for universes is based on dimensionality. Another important consideration is size. It appears that universes can have differing sizes and in fact can also grow or diminish in size (expansion or contraction). We will consider the size issue later. Other possible differentiating factors between universes will also be considered.

---

[80] Weinberg (1995) p. 216 exhibits an equation that relates the number of components of a spinor to the dimension of its space-time.

4.7 What is confined in a Universe?

The embedding of a universe within the Flatverse (eq. 1.4), neglecting quantum smearing which is presumably small relative to the size of a universe, implies a dividing surface between the contents of the universe and the "outside" Flatverse. The Flatverse is flat. It has no naked matter in it. It has no gravitation – otherwise it would not be flat.

So we conclude that the gravitational effects of mass-energy within a universe are confined to the universe. It is known that the total mass-energy of a universe is zero – the mass and energy of the contents are cancelled by the total gravitational energy.

Most particle features such as charge, and internal quantum numbers of a universe, also total to zero. Thus universes appear to confine almost everything within them assuming they can be embedded in a multiverse Flatverse. One exception, upon which we will dwell in chapter 6, is baryon number. Another exception is lepton number which is non-zero.

4.8 External Properties of Universes Within the Flatverse

Major non-null characteristics of our universe are its baryon number and its surface area. In the case of our universe the preponderance of matter over anti-matter yields an enormous baryon number. In chapter 6 we will consider the possibility of an abelian baryonic gauge field that would produce a force between universes, between a universe and baryons in another universe, and between baryons within a universe. This field, if it exists, and there is some evidence for it, would account for *baryon number conservation* and also would have significant consequences on the galactic level within our universe. Naturally it must be significantly weaker than the force of gravity. But the plenitude of baryons in our universe gives it importance on large scales.

There is also the possibility that a baryonic force (chapter 6) could induce movement, collisions, universe amalgamations, and other effects between galaxies in the multiverse making universes a peculiar form of particle on a multiverse scale. See chapters 5 and 6.

Another important characteristic of universes is their area. While a universe is not believed to be a black hole (although Hawking has recently jokingly suggested that our universe may be a black hole, and even more recently suggested black holes are not quite black holes – grey?), there are general qualitative similarities that lead us to consider the possibility that the four laws of black holes[81] may apply in part (or their entirety) to universes. In particular the 2nd law states

$$dM = \kappa dA/8\pi + \Omega dJ \tag{4.7}$$

where dM is the change in "mass/energy," A is the area of the Black Hole (universe), Ω is its angular velocity (zero for universes) and J is the angular momentum (also zero for universes). Since J = Ω = 0 for universes we can define a "mass" for a universe[82] in terms of a universe's area with

$$M = \kappa A/8\pi \tag{4.8}$$

This definition seems to capture the physics of universes that could be used in developing a dynamics of universes as we do in chapters 5 and 6. It also allows us to escape the dilemma of zero total energy of universes that would preclude treating universes as particles in the Flatverse and developing a Flatverse dynamics of universe-particles.

Thus we have universes with (large) baryon numbers and areas. Whether more complex universes exist with dimensions larger than 4 as suggested in section 4.6 is not known. We will assume that they do not exist for simplicity – again based on Leibniz's minimax principle.

4.9 The Surface of a Universe – A Type of Horizon

The surface of a universe may play a significant role in a fashion similar to the role of the surface of a black hole. Consider the non-quantum definition of an embedded universe defined by the sixteen equations:

$$z_i = f_i(x) \tag{1.4}$$

where the z_i are the 16 Flatverse coordinates, and x is a complex 4-vector in the universe. The metric is defined by

---

[81] Wald, R. M., "The Thermodynamics of Black Holes", *Living Reviews in Relativity* **4** (6): 12119 (2001).

[82] **We limit all subsequent discussions to spatially closed universes.**

$$g_{\mu\nu} = \partial f_j/\partial x^\mu \, \partial f_j/\partial x^\nu \tag{1.7}$$

A universe has a surface. Assuming rectangular coordinates one would expect that the surface[83] is approached when one or more of the coordinates x^μ approaches infinity where there is no matter and the metric approaches the flat space metric $\eta_{\mu\nu}$. Eq. 1.4 is valid only within the confines of the surface. The surface itself is defined by a set of Flatverse equations:

$$z_k = g_k(z) \tag{4.10}$$

Near the surface of a universe the Flatverse coordinates can be expressed in a simple rectangular form

$$\begin{aligned} z_1 &= f_1(x) \cong ix_0 \\ z_2 &= f_2(x) \cong x_1 \\ z_3 &= f_3(x) \cong x_2 \\ z_4 &= f_4(x) \cong x_3 \end{aligned} \tag{4.11}$$

which yields the metric

$$g_{\mu\nu} \cong \eta_{\mu\nu} \tag{4.12}$$

the flat space Lorentz metric.

The horizon (surface) of a universe thus encloses the universe in a manner that is quite similar to quark confinement in hadrons. Only now, in addition, to having total zero color SU(3) it also has total zero energy and a zero total value for all quantities except baryon number, lepton number and surface area. An amusing point of the horizon in comparison to quark confinement is that leakage of energy/matter through the horizon cannot happen. A universe has confinement. However it achieves confinement in a radically different way than quark confinement. A universe expands if energy/matter penetrates its horizon. So confinement is achieved by expansion. This confinement mechanism is not present in quark confinement.[84]

---

[83] Barring singularities within the universe and topological complications.

[84] Recently discrepancies have been found in the size of the proton in various experiments. These discrepancies are very small, and subject to dispute, but they are interesting when we consider "universe confinement."

Turning to Quantum Gravity: If we consider the behavior of the wave function of a universe at its horizon then we can only expect that it goes to zero as well in view of eq. 4.12. If universes collide then a quantum dynamical event occurs. The study of these phenomena experimentally is not technically feasible currently or in the near future.

4.10 Wormholes and Our Sister Universe

The existence of a sister universe, connected to our universe, has manifold advantages and supports the concept of a multiverse. The question of the mode of connection between these universes now arises. There are two evident possibilities:

1. The "so-called" sister universe could be part of our universe with the separation merely being the limitation of our perception to the four known dimensions. Speculations of this sort have appeared in the past. However, this possibility seems very unlikely since no decisive, fundamental reason for such a limitation of our experience of the "full" universe is apparent. The existence of such a hidden part would also raise serious questions of quantum entanglement as well as other issues.
2. The sister universe may be "connected" to our universe through a wormhole. The cover of this book displays a symbolic representation of two asymptotically flat universes connected by a wormhole (a closed universe). With the wormhole as the connecting link one can view our almost flat space-time as part of one asymptotically flat universe and our sister universe as another asymptotically flat universe. The separable equations that we considered in section 2.4 would apply to the asymptotically flat regions of both universes. (Thus their simple flat space-time form.) In the separation of the equations the locations in each separable part's components are very different. The wormhole itself plays no direct part in the equations. It merely connects the two universes through boundary conditions at the wormhole boundaries with the two asymptotically free universes.[85]

---

[85] See section II of Hawking, S. W., Phys. Rev. D **37** 904 (1988) for an example and section III for wormhole solutions of the Wheeler-DeWitt equation. A number of other authors, to which the reader is referred, have

The second possibility is the reasonable choice. It fulfills the need to connect to a sister universe in chapter 2. It should be noted that these universes, and their connecting wormhole, exist in the Flatverse.

The existence of a sister universe connected to our universe raises the question of their origin. In chapter 6 we suggest that the Flatverse contains an abelian gauge field that can give rise to these connected universes as a vacuum fluctuation.

4.11 Quantum Aspects of the Flatverse

The Flatverse contains quantized universes. Being quantized the horizon (surface) of a universe is not precisely defined but undergoes presumably mild quantum smearing. In particular the surface of a quantum universe in the Flatverse must be defined, as particle positions are defined in quantum mechanics, by a wave function whose "square" at each Flatverse point is the probability of a part of the universe being there. Physically we would expect the probability to fall sharply shortly beyond the classical horizon (surface) of the universe.

So in a Flatverse with a low density of universes most of the Flatverse will have zero probability of a universe part present.

There are however two other sources of quantum phenomena in the Flatverse: 1) an abelian baryonic gauge field that provides interactions between universes as well as quantum fluctuations that create universes; and 2) the Y field that appears within universes that quantizes the coordinates of each universe (See sections 1.12 and 1.13.). It also exists in the Flatverse to make abelian baryonic gauge field (chapter 6) perturbation theory computations finite to all order.

Gravitation is zero outside of universes. One can speculate that the abelian baryonic gauge field may be one of a set of gauge fields. Perhaps it includes a leptonic gauge field that would yield a lepton conservation law.

In any case we conclude that the Flatverse has a quantum field theoretic aspect. *We conjecture that most, if not all, of the universes in the Flatverse are quantum vacuum fluctuations of the baryonic gauge field.*

---

discussed the nature and role of wormholes. See Unruh, W. G., Phys. Rev. **D40**, 1053 (1989), and references therein.

4.12 Everett Universes Interpretation of Quantum Theory

Everett[86] has suggested that the many possible intermediate (measured) outcomes of a physical process cause its wave function to repeatedly fission to accommodate all possible outcomes at each stage of the physical process. This wave function fissioning corresponds physically to the repeated fission of the universes at each stage into new universes. The result, in Everett's view is an overwhelming cascading growth of universes for all physical processes in a given initial universe. This view has been adopted by a significant number of physicists and is often discussed.

We cannot agree with this interpretation of quantum mechanics. First because there exists no tangible evidence for this interpretation – nor is there likely to be evidence for it.[87] But most importantly because of the shear waste of the multiplication of universes which clearly contravenes the cleanness and economy of the features of our derivation of the Standard Model from geometry. It also contrasts with the directness of the proof of a sister universe that clarifies the origin of inertia and mass.[88] It also makes the Wheeler-DeWitt equation close to meaningless if universes are "continuously" bifurcating.

Perhaps the most significant conceptual defect of the Everitt interpretation is its violation of Leibniz's minimax principle by maximizing physical complexity in the multiplication of universes when the "minimalist" standard interpretation of quantum mechanics works so well in accounting for physical phenomena. The Everett interpretation leads to complexity without benefits.

---

[86] H. Everett, III, Rev. Mod. Phys. **29**, 454 (1957).

[87] What is not observable now, or in the future, is not physical.

[88] A sister universe would make the proposed fission of our universe more disastrous since the sister universe would have to also undergo fission through quantum entanglement. The combined fission process would seem to lead to major difficulties with the simplicity of quantum mechanics and quantum field theory as we see it in nature.

5. The Particle Interpretation of Universes

The Wheeler-DeWitt equation because of its similarity to the Klein-Gordon equation has led to numerous proposals to view universes as particles.[89]

In this chapter and the following two chapters we will speculate on a possible particle interpretation of universes that, while consistent with the spirit of the Wheeler-DeWitt equation and a Flatverse multiverse, goes far beyond our current experimental knowledge although some recent astronomical data tends to support it. It can only be justified in this century by its generality and simplicity. It just looks right.

The Wheeler-DeWitt equation specifies the internal dynamics of universes. The Flatverse and a universal baryonic gauge field embody the dynamics of universes. We view universes as extended particles in the 16-dimensional Flatverse rather like hadrons in particle physics with an unspecified internal structure. As we did in the low energy days of particle physics we will first quantize universes as point-like particles. We can then take account of their internal structure in interactions (between universes) using solutions of the Wheeler-DeWitt equation, spectral representations of vacuum expectation values, form factors, "deep inelastic" structure functions, and so on. The interaction between universes, and a universe and an elementary baryon particle in another universe, can be similarly treated.

The sole interaction between universes will be assumed to be an abelian baryonic gauge boson interaction. This interaction also occurs between a universe and a baryon particle, or between a pair of baryon particles. Gravity is not present in the Flatverse outside of universes. The baryonic gauge boson will be cloaked with Y fields through the use of Flatverse quantum coordinates as in sections 1.12 and 1.13 to avoid infinities.

---

[89] Some suggestions of this interpretation are: DeWitt, B. S., Phys. Rev. **160**, 1113 (1967); Robles-Perez, S. J., arXiv:1212.4598 (2012); and references therein.

5.1 The Hierarchy of the Cosmos

In our universe we have seen that natural phenomena form a hierarchy ranging from the simplest to the largest/most complex phenomena. One current view of the hierarchy of levels of physical phenomena is:

Elementary particles: leptons, quarks, gluons, gauge bosons, and Higgs particles
Hadrons: protons, neutrons, ...
Molecules
Agglomerations of molecules
Macroscopic objects
Planets
Stars
Galaxies
Clusters of galaxies
Supergalaxies
Universe

Each level generally has a set of "simplified" physical laws that describe its phenomena. For example molecules have quantum mechanical laws and regularities that help to understand the phenomena at the molecular level.

Interestingly, while all phenomena at each level should be explainable by the laws at lower levels, and ultimately all phenomena should be explainable at the level of elementary particles, connecting phenomena at different levels is often quite difficult and, in many cases, impossible.

Consequently, while we believe physical phenomena are ultimately reducible to the lowest level, the problem of relating phenomena at different levels is largely unresolved.

In this book we introduce new levels in the hierarchy of nature: the level of multiple universes, and the level of the all-encompassing Flatverse. In doing this, we seek to maintain what we know of our universe, as embodied in The New Standard Model and Quantum Gravity, while introducing one feature – an abelian baryon number gauge field, that appears to be needed to theoretically give a law of baryon number conservation as well as to provide a dynamics for universes in the Flatverse.[90]

---

[90] See Sakurai (1964) pp. 185-186 for early discussions. One might also argue for lepton number conservation (Sakurai (1964) pp. 190-194 provides an early discussion.) but we will forego that discussion in

We will now turn to a discussion of the universes level and a portrayal of universes as extended particles.

5.2 The Particle Interpretation of Extended Wheeler-DeWitt Equation Solutions

In section 4.5.2.1 we described features of the Wheeler-DeWitt equation that suggested that universes could be viewed as particles or anti-particles, or perhaps tachyons. The solutions of this equation are scalar wave functions on a manifold that are analogous to the solutions of the Klein-Gordon equation. Yet the issues of negative probabilities, possible tachyonic solutions, and negative frequency solutions suggest that an appropriate particle interpretation of universes is reasonable and can possibly resolve these problems.

Some physicists have taken the Wheeler-DeWitt equation as the starting point for a theory of a universe as a particle. The Wheeler-DeWitt equation describes the interior of a universe in a quantum framework. We will take a different approach using the Flatverse as the environment of universe particles. If the universes of the Flatverse are typically widely separated we can view a universe as an extended particle and begin by ignoring the detailed inner structure of universes. This approach is similar to the treatment of hadrons such as the proton as particles and developing a theory of them as fundamental particles using form factors, structure functions and so on to approximate their inner structure. Afterwards, as detailed data became available, the detailed investigation of the internal structure of hadrons using quark-parton models followed. We will pursue a similar theoretical development beginning with a "macroscopic" theory of universes as extended particles in the 16-dimension Flatverse. The internal structure of the particle universes will eventually be specified by the Wheeler-DeWitt equation expressed in Flatverse coordinates (section 4.3).

The two simplest choices for the nature of universes are "spin 0" universes and *fermionic* universes with odd half integer spin.[91] In this section we use Logic and spinor mathematics to establish a structural framework for

---

this work. Both conservation laws are still undergoing periodic testing with very low limits on possible violations. Universes also have excesses of leptons or anti-leptons that might be a source of dynamics via an abelian leptonic gauge field.

[91] Since the Flatverse is 16-dimensional, the spin of fermionic universe particles will be shown to be 127/2.

universes. We begin with the minimalist implication of Asynchronous Logic of (complex-valued) 4-dimensional universes resulting from a 4-valued logic.[92] Universes then initially become 4-dimensional logic constructs.

We will consider the possibility of fermionic universes and then briefly consider "spin 0" *bosonic* universes case later.

The first issue of fermionic universes (reminiscent of the discussions of spin in the 1920's) is the interpretation of spin states. We suggest that the upper (128) components (64 "spin up" and 64 "spin down") of a universe wave function represent a universe with an excess number of baryons. The lower (128) components lead to anti-universes where there is an excess of anti-baryons. These associations are analogous to the interpretations of the Dirac electron wave function.

The universe particle "spin up" and "spin down" states are distinguished by their interactions with the baryonic gauge field in a manner analogous to quantum electrodynamics. (The baryonic gauge field is described in the following chapter.)

5.3 "Free Field" Dynamics of Fermionic Universe Particles

We will view universes as extended particles with an odd half integer spin – *fermionic universe particles* - in the 16-dimensional Flatverse. In the Flatverse there are sixteen 256×256 matrices that are the equivalent of the four Dirac matrices in four dimensions. We will denote these sixteen matrices as γ^i for i = 1, 2, … , 16. They satisfy the anti-commutation relations:

$$\{\gamma^i, \gamma^j\} = 2\,\delta^{ij} \tag{5.1}$$

and thus form a Clifford algebra. We will choose y^{16} to be the time coordinate and thus imaginary (The 16-dimensional space is complex.) Therefore γ^{16} will be hermitean ($(\gamma^{16})^2 = 1$) and γ^i for i = 1, … , 15 will be anti-hermitean with $(\gamma^i)^2 = -1$. The number of linearly independent matrices in 16 dimensions is $2^{16} = 65{,}536$.

The Flatverse metric is

$$g^{ij} = -\delta^{ij}, \;\; g^{16} = 1 \tag{5.2}$$

---

92 Larger dimensioned universes are possible in the Flatverse. We will not consider these cases here.

for i, j = 1, 2, … , 15 and zero otherwise.

Except for the additional dimensions fermion dynamics is quite similar to the 4-dimensional case. The free universe particle Dirac equation is

$$(i\gamma^i\partial_i + m)\psi(y) = 0 \qquad (5.3)$$

summed over i = 1, 2, … , 16 where the mass is temporarily assumed to be constant and set by eq. 4.8. The derivative operator, is based on the use of quantum coordinates[93]

$$Y^i(y) = y^i + i\, Y_u^{\,i}(y)/M_u^{\,2} \qquad (5.4)$$

and is defined to be

$$\partial_i = \partial/\partial Y^i(y) = \partial/\partial(y_i - Y_{ui}(y)/M_u^{\,2}) \qquad (5.5)$$

where we assume $M_u = M_c$ with M_c being a very large mass scale of perhaps the order of the Planck mass.

$Y_u^{\,i}$ is a 16-dimensional Flatverse gauge field equivalent of $Y^\mu(x)$ used in Two Tier renormalization (discussed in chapter 1):

$$Y^\mu(z) = z^\mu + i\, Y^\mu(z)/M_u^{\,2} \qquad (1.35)$$

where $Y^\mu(z)$ is a free QED-like field. The $Y^i(y)$ quantum coordinates will be used in the Flatverse to eliminate potential divergences, in a manner similar to the case of our universe as outlined in section 1.12, when universe particle interactions are introduced later.

5.3.1 Four Types of Fermionic Universe Particles

There are four possible types of fermionic universe particles in the Flatverse that are similar to the four species of fermion described in section 1.10 (and Blaha (2010b)) of The New Standard Model. Two of these types are tachyonic. It is important to note that DeWitt points out that the Wheeler-DeWitt equation has tachyonic solutions since the mass-like term dependent on

[93] Giving Two Tier renormalization. See section 1.12.

${}^{(3)}R$ can be positive or negative.[94] A negative mass is an indication of tachyonic behavior wherein the wave propagation of the state functional is not necessarily in time-like directions and is thus tachyonic.

Eq. 5.3 is a type of 16-dimensional Dirac equation. There are three other general types of universe particle equations. The derivation of the four types of universe particles is similar to the derivation of fermion types in the Standard Model in 4-dimensional complex space-time given in Blaha (2010b).

The general form of a pure 16-dimensional complex Lorentz group[95] boost can be expressed in terms of a complex relative 15-velocity $\mathbf{v_c}$ between inertial reference frames. A 16-dimensional coordinate boost has the form

$$\Lambda_C(\mathbf{v_c}) \equiv \Lambda_C(\omega, \mathbf{v_c}) = \exp[i\omega\hat{\mathbf{w}}\cdot\mathbf{K}] \tag{5.6}$$

where

$$\omega = (\omega_r^{\,2} - \omega_i^{\,2} + 2i\omega_r\omega_i\, \hat{\mathbf{u}}_r\cdot\hat{\mathbf{u}}_i)^{1/2} \tag{5.7}$$

and

$$\hat{\mathbf{w}} = (\omega_r\hat{\mathbf{u}}_r + i\omega_i\hat{\mathbf{u}}_i)/\omega \tag{5.8}$$

with all vectors being 15-dimensional spatial vectors. We define the real and imaginary unit vectors $\hat{\mathbf{u}}_r\cdot\hat{\mathbf{u}}_r = 1 = \hat{\mathbf{u}}_i\cdot\hat{\mathbf{u}}_i$ with the result

$$\hat{\mathbf{w}}\cdot\hat{\mathbf{w}} = 1 \tag{5.9}$$

The complex relative velocity is

$$\mathbf{v_c} = \hat{\mathbf{w}}\tanh(\omega) \tag{5.10}$$

The free dynamical equations of the four universe particle species will be generated by 16-dimensional Lorentz boosts of the free Dirac equation of a universe particle at rest with the *requirement that the time variable* ($t = y^{16}$) *and energy is real in the resulting field equations.*[96] The procedure can most easily be performed in 16-dimensional momentum space and the Flatverse coordinate

---

[94] DeWitt, B. S., Phys. Rev. **160**, 1113 (1967) p. 1124.

[95] The 16-dimensional complex Lorentz group has similar features to the 4-dimensional complex Lorentz group. We shall only discuss it to the extent needed for our universe particle type's derivation. See Weinberg (1995) for the 4-dimensional Lorentz group – the 16-dimensional Lorentz group generalizes directly from the features of the 4-dimensional Lorentz group.

[96] The 16-dimensional "energy" must be real since it relates to the area of the universe – a real number.

space version of the generated equation determined from the momentum space version.

5.3.1.1 Dirac-like Equation – Type I universe Particle

A positive energy plane wave solution of the Dirac equation eq. 5.3 for a universe particle at rest is

$$\psi(y) = \exp[-imt]w(0) \tag{5.11}$$

where we set $\partial_t = \partial/\partial y_{16}$ while temporarily ignoring the $Y_u^{\,i}(y)/M_u^{\,2}$ term. w(0) is a 256 component spinor column vector. The solution $\psi(y)$ satisfies the momentum space Dirac equation for a particle at rest:

$$(m\gamma^{16} - m)\psi(y) = 0 \tag{5.12}$$

The 256 x 256 spinor matrix form of a 16-dimensional Lorentz boost with relative real velocity **v** of the Dirac matrices is[97]

$$S^{-1}(\Lambda(\mathbf{v}))\gamma^{\nu}S(\Lambda(\mathbf{v})) = \Lambda^{\nu}{}_{\mu}(\mathbf{v})\gamma^{\mu} \tag{5.13}$$

where $\Lambda^{\nu}{}_{\mu}(\mathbf{v})$ is a 16-dimensional Lorentz boost. $S(\Lambda(\mathbf{v}))$ has the form

$$S(\Lambda(\mathbf{v})) = \exp(-\omega\gamma^{16}\boldsymbol{\gamma}\cdot\mathbf{v}/(2|\mathbf{v}|))$$

$$= \cosh(\omega/2)I + \sinh(\omega/2)\gamma^{16}\boldsymbol{\gamma}\cdot\mathbf{p}/|\mathbf{p}| \tag{5.14}$$

with real ω = arctanh(|**v**|) and real **v.** |**p**| is the magnitude of the spatial 15-vector. Also

$$S^{-1}(\Lambda(\mathbf{v})) = \gamma^{16}S^{\dagger}(\Lambda(\mathbf{v}))\gamma^{16} = \exp(\omega\gamma^{16}\boldsymbol{\gamma}\cdot\mathbf{v}/(2|\mathbf{v}|))$$

$$= \cosh(\omega/2)I - \sinh(\omega/2)\gamma^{16}\boldsymbol{\gamma}\cdot\mathbf{p}/|\mathbf{p}| \tag{5.15}$$

---

[97] **The indices ν and μ from this point in this chapter have values: 1, 2, … , 16.**

If we now apply S(Λ(**v**)) to the momentum space Dirac equation of a particle at rest (eq. 5.12) we find

$$0 = S(\Lambda(\mathbf{v}))(m\gamma^{16} - m)\ \psi(y)$$
$$= [mS(\Lambda(\mathbf{v}))\gamma^{16}S^{-1}(\Lambda(\mathbf{v})) - m]S(\Lambda(\mathbf{v}))w(0)$$

A straightforward evaluation shows

$$mS(\Lambda(v))\gamma^{16}S^{-1}(\Lambda(v)) = g_{16\mu\nu}p^{\mu}\gamma^{\nu} = \not{p} \tag{5.16}$$

where p is a momentum 16-vector. In addition we define the 16-dimension spinor (256 components)

$$S(\Lambda(v))w(0) = w(p) \tag{5.17}$$

which can be viewed as a "positive energy Dirac spinor". The Dirac equation in momentum space has the familiar form:

$$(\not{p} - m)\ \exp[-ip\cdot y]w(p) = 0 \tag{5.18}$$

Eq. 5.18 implies the free, coordinate space Dirac equation:

$$(i\gamma^{\mu}\partial/\partial y^{\mu} - m)\psi(y) = 0 \tag{5.19}$$

We identify this equation as the dynamical equation of a type 1 universe particle. It corresponds to the free charged lepton elementary particle species.

5.3.1.2 Complex Boosts

The form of the 16-dimensional spinor boost transformation corresponding to the coordinate transformation eq. 5.6 is:

$$S_C(\omega, \mathbf{v}_c) \equiv S_C = \exp(-\omega\gamma^{16}\boldsymbol{\gamma}\cdot\hat{\mathbf{w}}/2)$$
$$= \cosh(\omega/2)I + \sinh(\omega/2)\gamma^{16}\boldsymbol{\gamma}\cdot\hat{\mathbf{w}} \tag{5.20}$$

with $\mathbf{v}_c$ and $\hat{\mathbf{w}}$ defined by eqs. 5.10 and 5.8 respectively. The inverse transformation is

$$S_C^{-1}(\omega, \mathbf{v_c}) = \exp(\omega\gamma^{16}\boldsymbol{\gamma}\cdot\hat{\mathbf{w}}/2)$$

$$= \cosh(\omega/2)I - \sinh(\omega/2)\gamma^{16}\boldsymbol{\gamma}\cdot\hat{\mathbf{w}} \qquad (5.21)$$

Note that S_C is not unitary just as in the 4-dimensional case.

We now apply a spinor boost to the Dirac equation for a particle at rest in this more general case of complex ω and $\hat{\mathbf{w}}$.

$$0 = S_C(\omega, \mathbf{v_c}))(m\gamma^{16} - m) \exp[-imt]w(0)$$
$$= [mS_C\gamma^{16}S_C^{-1} - m] \exp[-imt]S_C w(0) \qquad (5.22)$$

where $S_C = S_C(\omega, \mathbf{v_c})$. After some algebra we find

$$mS_C\gamma^{16}S_C^{-1} = m[\cosh(\omega)\gamma^{16} - \sinh(\omega)\boldsymbol{\gamma}\cdot\hat{\mathbf{w}}] \qquad (5.23)$$

We will use these complex boosts to generate the other species' Dirac-like equations.

5.3.1.3 Tachyon Universe particle Dirac Equation

The development of the spinor boost transformation (subsection 5.3.1.2 above) leads to two possible forms of the tachyon Dirac-like equation. One form will lead to a lagrangian dynamics for left-handed universe particles. The other form leads to a lagrangian dynamics for right-handed universe particles.

5.3.1.4 Type IIa Case: Left-Handed Tachyonic Universe Particles

If the real and imaginary relative vectors parts of $\hat{\mathbf{w}}$, namely $\hat{\mathbf{u}}_r$ and $\hat{\mathbf{u}}_i$, are parallel, then $\hat{\mathbf{u}}_r\cdot\hat{\mathbf{u}}_i = 1$ and

$$\omega = \omega_r + i\omega_i \qquad (5.24)$$

Eqs. 5.23 and 5.24 then imply

$$mS_C\gamma^{16}S_C^{-1} = m[\cosh(\omega_r)\cos(\omega_i) + i\sinh(\omega_r)\sin(\omega_i)]\gamma^{16} -$$
$$- m[\sinh(\omega_r)\cos(\omega_i) + i\cosh(\omega_r)\sin(\omega_i)]\boldsymbol{\gamma}\cdot\hat{\mathbf{u}}_r \qquad (5.25)$$

or

$$mS_C\gamma^{16}S_C^{-1} = \cos(\omega_i)\gamma\cdot p_r + i\sin(\omega_i)\gamma\cdot p_i \qquad (5.26)$$

where

$$p_r^{\mathbf{0}} = m\cosh(\omega_r) \qquad p_i^{\mathbf{0}} = m\sinh(\omega_r) \qquad (5.27)$$

and

$$\mathbf{p}_r = m\hat{\mathbf{u}}_r \sinh(\omega_r) \qquad \mathbf{p}_i = m\hat{\mathbf{u}}_r \cosh(\omega_r) \tag{5.28}$$

If $\omega_i = 0$, then we recover the momentum space Dirac-like equation. If $\omega_i = \pi/2$, then we obtain the left-handed momentum space tachyon equation:

$$mS_C\gamma^{16}S_C^{-1} = i\gamma{\cdot}p_i \tag{5.29}$$

and the tachyon energy and momentum expressions

$$\mathbf{p} = m\mathbf{v}\gamma_s \qquad E = m\gamma_s \tag{5.30}$$

where $\sinh(\omega) = \gamma_s = (\beta^2 - 1)^{-½}$ with $\beta = v/c > 1$. v is the absolute value of the 15 component spatial velocity. Also

$$S_C w(0) = w_C(p) \tag{5.31}$$

is a tachyon spinor.

The momentum space tachyonic Dirac-like equation is

$$(i\not{p} - m)\exp[-ip{\cdot}y]w_T(p) = 0 \tag{5.32}$$

where $p{\cdot}y = p^{16}y^{16} - \mathbf{p}{\cdot}\mathbf{y}$ after performing a corresponding boost in the exponential factor. If we apply $i\not{p}$ to eq. 5.32 we find the tachyon mass condition is satisfied

$$-E^2 + \mathbf{p}^2 = m^2 \tag{5.33}$$

Transforming back to coordinate space we obtain the "left-handed" *tachyonic Dirac-like equation*:

$$(\gamma^\mu\partial/\partial y^\mu - m)\psi_T(y) = 0 \tag{5.34}$$

5.3.1.5 Type IIb Case: Right-Handed Tachyonic Universe Particles

If the real and imaginary relative vectors parts of $\hat{\mathbf{w}}$, $\hat{\mathbf{u}}_r$ and $\hat{\mathbf{u}}_i$, are anti-parallel $\hat{\mathbf{u}}_r = -\hat{\mathbf{u}}_i$, then $\hat{\mathbf{u}}_r{\cdot}\hat{\mathbf{u}}_i = -1$ and

$$\omega = \omega_r - i\omega_i \tag{5.35}$$

then

$$mS_C\gamma^{16}S_C^{-1} = m[\cosh(\omega_r)\cos(\omega_i) - i\sinh(\omega_r)\sin(\omega_i)]\gamma^{16} - \\ - m[\sinh(\omega_r)\cos(\omega_i) - i\cosh(\omega_r)\sin(\omega_i)]\boldsymbol{\gamma}\cdot\hat{\mathbf{u}}_r \tag{5.36}$$

or

$$mS_C\gamma^{16}S_C^{-1} = \cos(\omega_i)\gamma\cdot p_r - i\sin(\omega_i)\gamma\cdot p_i \tag{5.37}$$

where

$$p_r^{16} = m\cosh(\omega_r) \qquad p_i^{16} = m\sinh(\omega_r) \tag{5.38}$$

and

$$\mathbf{p}_r = m\hat{\mathbf{u}}_r\sinh(\omega_r) \qquad \mathbf{p}_i = m\hat{\mathbf{u}}_r\cosh(\omega_r) \tag{5.39}$$

If $\omega_i = \pi/2$, then we obtain the right-handed momentum space tachyon equation.[98]

$$(-\gamma^\mu\partial/\partial y^\mu - m)\psi_T(y) = 0 \tag{5.40}$$

5.3.1.6 Type III Case: "Up-Quark-like" Universe Particles

There are two other cases where we can obtain fermion dynamical equations with a *real* time variable and real energy. In one case we set $\hat{\mathbf{u}}_r\cdot\hat{\mathbf{u}}_i = 0$ and have a real ω.

If the real and imaginary relative vectors parts of $\hat{\mathbf{w}}$, namely $\hat{\mathbf{u}}_r$ and $\hat{\mathbf{u}}_i$, are perpendicular, $\hat{\mathbf{u}}_r\cdot\hat{\mathbf{u}}_i = 0$, then

$$\omega = (\omega_r^2 - \omega_i^2)^{1/2} \tag{5.41}$$

Thus ω is either pure real ($\omega_r \geq \omega_i$) or pure imaginary ($\omega_r < \omega_i$).

The momentum space equation generated by the corresponding spinor boost is

$$\{m\cosh(\omega)\gamma^{16} - m\sinh(\omega)\boldsymbol{\gamma}\cdot(\omega_r\hat{\mathbf{u}}_r + i\omega_i\hat{\mathbf{u}}_i)/\omega - m\}\exp[-imt]w_c(p) = 0 \tag{5.42}$$

Defining the momentum 4-vector

---

[98] We note that $\gamma_s = (\beta^2 - 1)^{-1/2}$, if expressed in terms of ω, has a branch cut extending from <-∞, +∞> in the complex ω plane. Thus values of ω with positive imaginary parts are physically different from values of ω with negative imaginary parts.

$$p = (p^{16}, \mathbf{p}) \tag{5.43}$$

where

$$p^{16} = m \cosh(\omega) \qquad \mathbf{p} = \mathbf{p}_r + i\mathbf{p}_i \tag{5.44}$$

with

$$\mathbf{p}_r = m\omega_r\hat{\mathbf{u}}_r \sinh(\omega)/\omega \qquad \mathbf{p}_i = m\omega_i\hat{\mathbf{u}}_i \sinh(\omega)/\omega \tag{5.45}$$

$$\mathbf{p}_r \cdot \mathbf{p}_i = 0 \tag{5.46}$$

then we obtain a positive energy Dirac-like equation

$$[p\cdot\gamma - m]\exp[-imt]w_c(p) = 0$$

or

$$[p^{16}\gamma^{16} - (\mathbf{p}_r + i\mathbf{p}_i)\cdot\boldsymbol{\gamma} - m]\exp[-ip\cdot y]w_c(p) = 0 \tag{5.47}$$

with a complex 3-momentum **p** and the 4-momentum mass shell condition:

$$p^2 = p^{16\,2} - \mathbf{p}_r\cdot\mathbf{p}_r + \mathbf{p}_i\cdot\mathbf{p}_i = m^2 \tag{5.48}$$

Note

$$|\mathbf{v}_c| = |\mathbf{p}|/p^{16} = [(\mathbf{p}_r + i\mathbf{p}_i)\cdot(\mathbf{p}_r + i\mathbf{p}_i)]^{1/2}/p^{16} = \tanh(\omega) \tag{5.49}$$

and so the Lorentz factor is

$$\gamma = \cosh(\omega) \tag{5.50}$$

Eq. 5.47 is the momentum space equivalent of the wave equation[99]

$$[i\gamma^{16}\partial/\partial t + i\boldsymbol{\gamma}\cdot(\nabla_r + i\nabla_i) - m]\psi_u(t, \mathbf{y}_r, \mathbf{y}_i) = 0 \tag{5.51}$$

where $\mathbf{y} = \mathbf{y}_r - i\mathbf{y}_i$, and where the grad operators ∇_r and ∇_i are with respect to $\mathbf{y}_r$ and $\mathbf{y}_i$ respectively. Since $\hat{\mathbf{u}}_r\cdot\hat{\mathbf{u}}_i = 0$ we see that there is a subsidiary condition on the wave function

$$\nabla_r\cdot\nabla_i\, \psi_u(t, \mathbf{y}_r, \mathbf{y}_i) = 0 \tag{5.52}$$

---

[99] The gradient operators ∇_r and ∇_i are 15-dimensional spatial gradient operators.

We note eq. 5.52 can be put into covariant form as the difference of two vectors squared (which is a real 16-dimensional Lorentz group invariant):

$$[\gamma^{16}\partial/\partial t + i\boldsymbol{\gamma}\cdot(\nabla_r + i\nabla_i)]^2 - [\gamma^{16}\partial/\partial t + i\boldsymbol{\gamma}\cdot(\nabla_r - i\nabla_i)]^2 = 4\nabla_r\cdot\nabla_i.$$

We identify eq. 5.51 as the dynamical equation of an "up-quark-like" universe particle.

5.3.1.7 Type IVa Case: Left-Handed "Down-Quark-like" Tachyonic Universe Particles

In this case we set $\hat{\mathbf{u}}_r\cdot\hat{\mathbf{u}}_i = 0$. Then by eq. 5.7

$$\omega = (\omega_r^2 - \omega_i^2)^{1/2}$$

Thus ω again starts out either pure real (if $\omega_r \geq \omega_i$) or pure imaginary (if $\omega_r < \omega_i$). In this case we also choose ω real, and then change ω to

$$\omega = (\omega_r^2 - \omega_i^2)^{1/2} \rightarrow \omega' = (\omega_r^2 - \omega_i^2)^{1/2} + i\pi/2 = \omega + i\pi/2$$

by adding $i\pi/2$ to ω since ω is a free parameter. We then proceed as we did in the prior tachyon case.[100]. The resulting Lorentz boost

$$\Lambda_C = \exp[i((\omega_r^2 - \omega_i^2)^{1/2} + i\pi/2)(\omega_r\hat{\mathbf{u}}_r + i\omega_i\hat{\mathbf{u}}_i)\cdot\mathbf{K}/\omega] \qquad (5.53)$$

becomes a left-handed "quark-like" boost. The tachyon dynamical equation is[101]

$$[\gamma^{16}\partial/\partial t + \boldsymbol{\gamma}\cdot(\nabla_r + i\nabla_i) - m]\psi_d(y) = 0 \qquad (5.54)$$

with the constraint equation

$$\nabla_r\cdot\nabla_i\,\psi_d(t, \mathbf{y}_r, \mathbf{y}_i) = 0 \qquad (5.55)$$

---

[100] Here again the choice of ω in eq. 5.53 leads to a "left-handed" universe particle while the choice $\omega' = \omega - i\pi/2$ leads to a right-handed one.

[101] The gradient operators ∇_r and ∇_i are 15-dimensional spatial gradient operators.

We will call the universe particles satisfying eqs. 5.54 and 5.55 left-handed *tachyonic quark-like universe particles.*

5.3.1.8 Type IVb Case: Right-Handed Down-Quark-like Tachyonic Universe Particles

In this case we set $\hat{\mathbf{u}}_r \cdot \hat{\mathbf{u}}_i = 0$. Then by eq. 5.7

$$\omega = (\omega_r^2 - \omega_i^2)^{1/2}$$

Thus ω again starts out either pure real (if $\omega_r \geq \omega_i$) or pure imaginary (if $\omega_r < \omega_i$). In this case we also choose ω real, and then change ω to

$$\omega = (\omega_r^2 - \omega_i^2)^{1/2} \rightarrow \omega' = (\omega_r^2 - \omega_i^2)^{1/2} - i\pi/2 = \omega - i\pi/2$$

since ω is a free parameter and proceed as we did in the prior case. The resulting Lorentz boost

$$\Lambda_C = \exp[i((\omega_r^2 - \omega_i^2)^{1/2} - i\pi/2)(\omega_r\hat{\mathbf{u}}_r + i\omega_i\hat{\mathbf{u}}_i)\cdot\mathbf{K}/\omega] \tag{5.56}$$

becomes a right-handed quark-like boost. The resulting tachyon dynamical equation is

$$[-\gamma^{16}\partial/\partial t - \boldsymbol{\gamma}\cdot(\nabla_r + i\nabla_i) - m]\psi_d(y) = 0 \tag{5.57}$$

with the constraint equation

$$\nabla_r\cdot\nabla_i\,\psi_d(t, \mathbf{y}_r, \mathbf{y}_i) = 0 \tag{5.58}$$

We will call the universe particles satisfying eqs. 5.57 and 5.58 right-handed *tachyonic quark-like universe particles.*

5.3.2 Lagrangians

In this section we will develop a lagrangian formalism for each of the four types of universe particles noting that a tachyonic universe particles have two forms: left-handed and right-handed (discussed later in section 5.3.5).

The various types of universe particles described in section 5.3.1 correspond to universes with differing internal characteristics and motion in the Flatverse. The equations are all free field equations. Internal potentials and interactions must be introduced in these equations to complete the universe dynamical equations. A connection to the Wheeler-DeWitt description of their internal quantum structure also remains to be established (section 5.3.6).

In defining the lagrangians for the four universe types that yield their dynamical equations in a canonical manner, we require the conventional quantum field theory feature that the hamiltonian derived from the lagrangian is hermitean. We will develop a separate lagrangian for each type.

5.3.2.1 Type I Universe Particle Lagrangian

The Universe particle Dirac equation lagrangian is

$$\mathcal{L}_u = \bar{\psi}(i\gamma^\mu \partial/\partial y^\mu - m)\psi(y) \tag{5.59}$$

where

$$\bar{\psi} = \psi^\dagger \gamma^{16}$$

and $\psi^\dagger$ is the hermitean conjugate of ψ.

5.3.2.2 Type II Tachyon Universe Particle Lagrangian

This lagrangian includes both left-handed and right-handed cases. It can be separated into lagrangian terms for each case using parity projection operators.

$$\mathcal{L}_{uT} = \psi_T^{\,S}(\gamma^\mu \partial/\partial y^\mu - m)\psi_T(y) \tag{5.60}$$

where

$$\psi_T^{\,S} = \psi_T^{\,\dagger}\, i\gamma^{16}\gamma^5 \tag{5.61}$$

The peculiar form of the tachyon universe lagrangian is necessitated by the hermiticity of the hamiltonian calculated from it.

5.3.2.3 Type III "Up-Quark-like" Universe Particle Lagrangian

The lagrangian density of a free "up-quark-like" universe particle is

$$\mathcal{L}_u = \bar{\psi}_u(i\gamma^\mu D_\mu - m)\psi_u(y) \tag{5.62}$$

where $\bar{\psi}_u = \psi_u^{\dagger}\gamma^{16}$ and

$$\psi_u^{\dagger} = [\psi_u(\mathbf{y}_r, \mathbf{y}_i)]^{\dagger} \,|_{\mathbf{y}_i = -\mathbf{y}_i} \tag{5.63}$$

$$D_{16} = \partial/\partial y^{16}$$
$$D_k = \partial/\partial y_r^{\,k} + i\,\partial/\partial y_i^{\,k} \tag{5.64}$$

for k = 1, 2, … , 15. The action

$$I = \int d^{15}y \mathcal{L}_u \tag{5.65}$$

It is easy to show that this action is also real.

5.3.2.4 Type IV "Down-Quark-like" Tachyon Universe Particle Lagrangian

The lagrangian density of a free "down-quark-like" universe particle is

$$\mathcal{L}_d = \psi_d^{\,C}(y)(\gamma^{16}\partial/\partial t + \boldsymbol{\gamma}\cdot(\nabla_r + i\nabla_i) - m)\psi_d(y) \tag{5.66}$$

where

$$\psi_d^{\,C}(y) = [\psi_d(y)]^{\dagger}\,|_{\mathbf{y}_i = -\mathbf{y}_i}\, i\gamma^{16}\gamma^5 \tag{5.67}$$

In words, eq. 5.67 states: take the hermitean conjugate of $\psi_d(y)$; change $\mathbf{y}_i$ to $-\mathbf{y}_i$; and then post-multiply by the indicated factors.

The action is

$$I = \int d^{15}y \mathcal{L}_d \tag{5.68}$$

The action is real. The lagrangian can also be separated into left-handed and right-handed parts using projection operators.

5.3.3 Form of The Flatverse Quantum Coordinates Gauge Field

The discussions of sections 5.3.1 and 5.3.2 assumed the coordinates were Flatverse coordinates and their derivatives. Prior to those discussions we indicated we would use quantum coordinates in the Flatverse of the form[102]

$$Y^i(y) = y^i + i\, Y_u^{\,i}(y)/M_u^{\,8} \qquad (5.4)$$

and their derivatives

$$\partial_i = \partial/\partial Y^i(y) = \partial/\partial(y^i - Y_u^{\,i}(y)/M_u^{\,8}) \qquad (5.5)$$

for i = 1, 2, ... , 16 to eliminate divergences in quantum field theory. The subscript "u" signifies universes. The mass constant for the Flatverse M_u may be the same as the mass constant M_c appearing in the Two Tier mechanism for our universe. (See section 1.12 for a discussion of eliminating infinities with this mechanism.)

In this section we define the gauge fields $Y_u^{\,i}(y)$ of the Flatverse. They are similar to the $Y^\mu(y)$ fields of our New Standard Model.[103] The $Y_u(y)$ 16-dimensional vector gauge field, in the absence of external sources, will be defined in a 16-dimensional Coulomb gauge:

$$Y_u^{\,16}(y) = 0 \qquad (5.69)$$
$$\partial\, Y_u^{\,j}(y)/\partial y^j = 0$$

where the sum over j is over the 15 spatial y coordinates. We follow a procedure similar to Blaha (2003) but for 16-dimensional space. The lagrangian density for the free $Y_u^{\,j}(y)$ fields is

$$\mathscr{L}_u = -\tfrac{1}{4}\, F_u^{\,\mu\nu} F_{u\mu\nu} \qquad (5.70)$$

and the lagrangian is

$$L_u = \int d^{15}y\, \mathscr{L}_u \qquad (5.71a)$$

with

$$F_{u\mu\nu} = \partial Y_{u\mu}/\partial y^\nu - \partial Y_{u\nu}/\partial y^\mu \qquad (5.71b)$$

---

[102] The denominator $M_u^{\,8}$ is necessitated by the dimension of $Y_u^{\,i}(y)$ which is $[m]^7$. Eqs. 5.78 and 5.81 below imply this conclusion.

[103] See Blaha (2005a) for details.

The equal time commutation relations, derived in the usual way, are:

$$[Y_u^{\mu}(\mathbf{y}, y^0), Y_u^{\nu}(\mathbf{y}', y^0)] = [\pi_u^{\mu}(\mathbf{y}, y^0), \pi_u^{\nu}(\mathbf{y}', y^0)] = 0 \quad (5.72)$$

$$[\pi_u^{j}(\mathbf{y}, y^0), Y_{uk}(\mathbf{y}', y^0)] = -i\,\delta^{15tr}{}_{jk}(\mathbf{y} - \mathbf{y}') \quad (5.73)$$

for μ, ν, j, k = 1, 2, … , 15 where

$$\pi_u^{k} = \partial \mathscr{L}_u / \partial Y_{uk}' \quad (5.74)$$

$$\pi_u^{0} = 0 \quad (5.75)$$

and

$$\delta^{tr}{}_{jk}(\mathbf{y} - \mathbf{y}') = \int d^{15}k\, e^{i\,\mathbf{k}\bullet(\mathbf{y}-\mathbf{y}')}\,(\delta_{jk} - k_j k_k/\mathbf{k}^2)/(2\pi)^{15} \quad (5.76)$$

$$Y_{uk}' = \partial Y_{uk}/\partial y^{16} \quad (5.77)$$

The Coulomb gauge indicates fourteen degrees of freedom are present in the vector potential. The Fourier expansion of the vector potential is:

$$Y_u^{i}(y) = \int d^{15}k\, N_0(k) \sum_{\lambda=1}^{14} \varepsilon^{i}(k, \lambda)[a(k,\lambda)\, e^{-ik\cdot y} + a^{\dagger}(k,\lambda)\, e^{ik\cdot y}] \quad (5.78)$$

where

$$N_0(k) = [(2\pi)^{15} 2\omega_k]^{-1/2} \quad (5.79)$$

and (since the field is massless)

$$k^{16} = \omega_k = (\mathbf{k}^2)^{1/2} \quad (5.80)$$

where k^{16} is the energy, and where the $\varepsilon^{i}(k, \lambda)$ are the polarization unit vectors for $\lambda = 1, \dots, 14$ and $k^{\mu}k_{\mu} = k^{16\,2} - \mathbf{k}^2 = 0$.

The commutation relations of the Fourier coefficient operators are:

$$[a(k,\lambda), a^{\dagger}(k',\lambda')] = \delta_{\lambda\lambda'}\delta^{15}(\mathbf{k} - \mathbf{k}') \quad (5.81)$$

$$[a^{\dagger}(k,\lambda), a^{\dagger}(k',\lambda')] = [a(k,\lambda), a(k',\lambda')] = 0 \quad (5.82)$$

and the polarization vectors satisfy

$$\sum_{\lambda=1}^{14} \varepsilon_i(k, \lambda)\varepsilon_j(k, \lambda) = (\delta_{ij} - k_i k_j/\mathbf{k}^2) \quad (5.83)$$

It will be convenient to divide the Y field into positive and negative frequency parts:

$$Y_u{}^+{}_i(y) = \int d^{15}k\ N_0(k) \sum_{\lambda=1}^{14} \varepsilon_i(k, \lambda)\ a(k,\lambda)\ e^{-ik\cdot y} \qquad (5.84)$$

and

$$Y_u{}^-{}_i(y) = \int d^{15}k\ N_0(k) \sum_{\lambda=1}^{14} \varepsilon_i(k, \lambda)\ a^\dagger(k,\lambda)\ e^{ik\cdot y} \qquad (5.85)$$

For later use we note the commutator between the positive and negative frequency parts is:

$$[\ Y_u{}^-{}_j(y_1),\ Y_u{}^+{}_k(y_2)] = -\int d^{15}k\ e^{ik\cdot(y_1 - y_2)}\ (\delta_{jk} - k_j k_k/\mathbf{k}^2)/[(2\pi)^{15} 2\omega_k] \qquad (5.86)$$

5.3.3.1 Y_u Fock Space Imaginary Coordinate States

States can also be defines for the quantized Y_u field. These states will be similar in form to electromagnetic photon states but play a different role in our approach since they are in fact coordinate excitation states for the imaginary part of $Y^i(y)$ (eq. 5.4). Thus universe particles (and other fields) will exist in a real 16-dimensional space with quantum excitations into imaginary Quantum Dimensions. These excitations become significant at high energies. At the low energies space appears c-number complex; at very high energies space becomes slightly q-number complex.

There are two types of imaginary coordinate excitations: 1.) Quantum excitations into Fock states consisting of a superposition of states with a definite finite number of Y_u "particles" and 2.) Imaginary coordinate excitations into coherent Y_u states with an "infinite" number of particles. Coherent states can be viewed as representing "classical" fields.

In this section we will consider Y_u field states with a definite number of excitations ("particles"). The raising and lowering operators of the Y_u field can be used to define free particle states. For example a one particle state can be defined by

$$|k, \lambda\rangle = a^\dagger(k, \lambda)|0\rangle \qquad (5.87)$$

with corresponding bra state

$$<k, \lambda| = <0|a(k, \lambda) \quad (5.88)$$

where the "coordinate vacuum" is defined as usual:

$$a(k, \lambda)|0> = 0 \quad (5.89)$$

$$<0|a^\dagger(k, \lambda) = 0 \quad (5.90)$$

Multi-particle states can also be defined in the conventional way with products of the raising and lowering operators applied to the vacuum. The set of all states containing a finite number of "particles" constitutes a Fock space.

A state with a finite number of Y_u "particles" represents a quantum fluctuation into imaginary Quantum Dimensions.

5.3.3.2 Y_u Coherent Imaginary Coordinate States

Coherent Y_u states bring us closer what we might consider to be "classical" imaginary dimensions – dimensions that we can, in principle, experience as we do normal dimensions. Let us define the coherent state[104]

$$| y, p> = e^{-\mathbf{p}\cdot \mathbf{Y}_u^-(y)/M_u^8}|0> \quad (5.91)$$

This state is an eigenstate of the coordinate operator $Y_u^+(y')$:

$$Y_u{}^+{}_j(y_1)\ |y_2, p> = -[Y_u{}^+{}_j(y_1), \mathbf{p}\cdot\mathbf{Y}^-(y_2)]/M_u^8|y, p> \quad (5.92)$$

$$= -\int d^{15}k\ [N_0(k)]^2\ e^{ik\cdot(y_2-y_1)}\ (p_j - k_j\mathbf{p}\cdot\mathbf{k}/\mathbf{k}^2)/M_u^8|y, p>$$

$$= p^i\Delta_{Tij}(y_1 - y_2)/M_u^8|y, p> \quad (5.93)$$

where $p^i\Delta_{Tij}(y_1 - y_2)/M_u^8$ is the eigenvalue of $Y_u{}^+{}_j(y_1)$. As we will see later, the eigenvalue of Y_u^+ becomes large as $(y_1 - y_2)^2 \to 0$. Thus the imaginary Quantum

[104] Coherent states are well known in the physics literature. See for example T. W. B. Kibble, J. Math. Phys. **9**, 315 (1968) and references therein; V. Chung, Phys. Rev. **140**, B1110 (1965); J. R. Klauder, J. McKenna, and E. J. Woods, J. Math. Phys. **7**, 822 (1966) and references therein.

Dimensions become significant at very short distances, and significantly modify the high-energy behavior of quantum field theories. In particular, Quantum Dimensions have a significant effect when

$$(y_1 - y_2)^2 \lessapprox (2^{14}\pi^{14}M_u^{\,2})^{-1} \qquad (5.94)$$

We assume the mass scale M_u is very large – perhaps of the order of the Planck mass (1.221×10^{19} GeV/c^2).

5.3.3.3 Quantization of the Type I Free Universe Particle Dirac Field

The quantization procedure is formally identical to that of a conventional Dirac particle. The standard equal time anti-commutation relations for a 16-dimensional fermion field are:

$$\{\psi_a(Y), \psi_\beta(Y')\} = \{\pi_{\psi a}(Y), \pi_{\psi\beta}(Y')\} = 0 \qquad (5.95)$$

$$\{\pi_{\psi a}(Y), \psi_\beta(Y')\} = i\,\delta_{a\beta}\,\delta^{15}(\mathbf{Y} - \mathbf{Y'}) \qquad (5.96)$$

where a and β are the spinor indices ranging from 1 to 256 and where

$$\pi_{\psi a}(Y) = i\,\psi_a^{\dagger}(Y) \qquad (5.97)$$

The field can be expanded in a fourier series:

$$\psi(Y(y)) = \sum_s \int d^{15}p\; N^d{}_m(p)\,[b(p,s)u(p,s) :e^{-ip\cdot(y + iYu/M_u^{\,8})}: +$$
$$+ d^{\dagger}(p,s)v(p,s) :e^{ip\cdot(y + iYu/M_u^{\,8})}:] \qquad (5.98)$$

$$\psi^{\dagger}(Y(y)) = \sum_s \int d^{15}p\; N^d{}_m(p)\,[b^{\dagger}(p,s)\bar{u}(p,s)\gamma^0 :e^{+ip\cdot(y + iYu/M_u^{\,8})}: +$$
$$+ d(p,s)\bar{v}(p,s)\gamma^0 :e^{-ip\cdot(y + iYu/M_u^{\,8})}:] \qquad (5.99)$$

where

$$N^d{}_m(p) = [m/((2\pi)^{15}E_p)]^{1/2} \qquad (5.100)$$

and

$$E_p = p^{16} = (\mathbf{p}^2 + m^2)^{1/2} \qquad (5.101)$$

The commutation relations of the Fourier coefficient operators are:

$$\{b(p,s), b^\dagger(p',s')\} = \delta_{ss'}\delta^{15}(\mathbf{p} - \mathbf{p}') \quad (5.102)$$
$$\{d(p,s), d^\dagger(p',s')\} = \delta_{ss'}\delta^{15}(\mathbf{p} - \mathbf{p}') \quad (5.103)$$
$$\{b(p,s), b(p',s')\} = \{d(p,s), d(p',s')\} = 0 \quad (5.104)$$
$$\{b^\dagger(p,s), b^\dagger(p',s')\} = \{d^\dagger(p,s), d^\dagger(p',s')\} = 0 \quad (5.105)$$
$$\{b(p,s), d^\dagger(p',s')\} = \{d(p,s), b^\dagger(p',s')\} = 0 \quad (5.106)$$
$$\{b^\dagger(p,s), d^\dagger(p',s')\} = \{d(p,s), b(p',s')\} = 0 \quad (5.107)$$

The spinors u(p,s) and v(p,s) are defined in a conventional way (as in Kaku, and in Bjorken and Drell). However their form is different from the 4-dimensional case. If one takes the 256×256 $\gamma\cdot p$ matrix, then the first 128 columns give u(p,s) up to a normalization for the free particle case, the remaining 128 columns give v(p,s) up to a normalization.

Since there are 256 possible spin values, using the equation 2s + 1 = total number of spin values we see that the spin of a fermionic universe particle is s = 127/2. The possible universe particle spin values are:

Up spin values: +1/128, +2/128, … , +64/128 = ½
Down spin values: -64/128 = -½, -63/128, … , -1/128

5.3.3.4 Feynman Propagators for the Type I Free Universe Particle Dirac Field

The form of the fermionic universe particle Feynman propagator differs from a conventional fermion propagator by having a Gaussian factor R(**p**, z) in its fourier expansions. This follows from using quantum Flatverse coordinates (eq. 5.4).

$$iS_F^{TT}(y_1 - y_2) = \langle 0|T(\bar{\psi}(Y(y_1))\psi(Y(y_2)))|0\rangle \quad (5.108)$$

where the time ordering is with respect to y_1^{16} and y_2^{16}. Expanding the free fields leads to the fourier representation:

$$iS_F^{TT}(y_1 - y_2) = i\int \frac{d^{16}p\, e^{-ip\cdot(y_1 - y_2)}\,(\not{p} + m)\,R(\mathbf{p}, y_1 - y_2)}{(2\pi)^{16}\,(p^2 - m^2 + i\varepsilon)} \quad (5.109)$$

where

$$R(\mathbf{p}, y_1 - y_2) = \exp[-p^ip^j\Delta_{Tij}(y_1 - y_2)/M_u^{16}] \quad (5.110)$$

$$= \exp\{-p^2[A(v) + B(v)\cos^2\theta] / [(2\pi)^{14}M_c^{\ 4}z^2]\} \qquad (5.111)$$

with

$$z^\mu = y_1^{\ \mu} - y_2^{\ \mu} \qquad (5.112)$$
$$z = |\mathbf{z}| = |\mathbf{y}_1 - \mathbf{y}_2| \qquad (5.113)$$
$$p = |\mathbf{p}| \qquad (5.114)$$
$$v = |z^0|/z \qquad (5.115)$$
$$A(v) = (1 - v^2)^{-1} + .5v \ln[(v - 1)/(v + 1)] \qquad (5.116)$$
$$B(v) = v^2(1 - v^2)^{-1} - 1.5v \ln[(v - 1)/(v + 1)] \qquad (5.117)$$
$$\mathbf{p}\cdot\mathbf{z} = pz\cos\theta \qquad (5.118)$$

and $|\mathbf{p}|$ denoting the length of a spatial 15-vector $\mathbf{p}$ while $|z^0|$ is the absolute value of $z^0 \equiv z^{16}$.

As eq. 5.109 indicates, the Gaussian damping factor R(p, z) for large spatial momentum p is the same for both the positive and negative frequency parts of the two-tier Feynman propagator. We are assuming the spatial momentum is real-valued in this discussion. It is also important to note that R(p, z) does not depend on $p^0 = p^{16}$ (in the Y Coulomb gauge) and thus the integration over p^0 proceeds in the usual way to produce time-ordered positive and negative frequency parts.

5.3.3.5 Feynman Propagators for the Types II, III, and IV Free Universe Particle Dirac Fields

These propagators differ in details from the Type I propagator. The differences modulo the change in dimension appear in Blaha (2011c). See also Blaha (2005a) for a detailed discussion of 4-dimensional spin ½ particle propagators.

5.3.4 Expanding and Contracting Universes: Impact of Time Dependent Universe Particle Masses

Our discussions of the dynamics of universe particles assumed their masses were constant. However the definition of mass in terms of the area of a universe based on the physics of black holes is

$$M = \kappa A/8\pi \qquad (4.8)$$

where A is the area of the black hole shows that the mass of a universe particle is time dependent because the area of a universe is time dependent. For example, our universe is expanding and its surface area is thus growing with time.

Eqs. 5.11 (and subsequent fermionic dynamic equations) must then be modified from

$$\psi(y) = \exp[-imt]w(0) \tag{5.11}$$

to a covariant form:

$$\psi(y) = \exp[-i\int_0^{w\cdot y} m(t')dt']w(0) \tag{5.119}$$

where w is a unit 16-vector in the time (y^{16}) direction ($w^2 = 1$). The lower bound on the integral, 0, is the time of the beginning of the universe particle – its Big Bang. Thus the cumulative change in the mass of the universe particle is significant. It is interesting to note that the Wheeler-Dewitt equation (eq. 3.37) also has a variable value mass term R that also depends on the evolution of the universe.

Eq. 5.119 satisfies the free covariant Dirac-like universe particle field dynamic equation

$$[i\gamma^i\partial/\partial y^i - m(w\cdot y)]\psi(y) = 0 \tag{5.120}$$

in contrast to the constant mass equation eq. 5.19. Substituting eq. 5.119 in eq. 5.120 we find

$$(\gamma^i w_i\, m(w\cdot y) - m(w\cdot y))\psi(y) = 0 \tag{5.121}$$

or

$$(\gamma^i w_i - 1)\psi(y) = 0 \tag{5.122}$$

Upon performing a 16-dimensional Lorentz boost (of the type of eqs. 5.13 – 5.16) on eq. 5.122 we obtain

$$(\gamma_i p^i/m_0 - 1)\psi(y) = 0$$

or

$$(\gamma_i p^i - m_0)\psi(y) = 0 \tag{5.123}$$

with

$$\psi(y) = \exp[-i\int_0^{p\cdot y/m_0} m(t')dt']w(p) \tag{5.124}$$

where p^i is a momentum 16-vector with $p^2 = m_0^2$. Eq. 5.123 is the constant mass momentum space dynamic equation it determines the spinor in $\psi(y)$. After taking account of the quantum coordinates the quantum Dirac-like universe particle wave function has the form

$$\psi(Y(y)) = \sum_s \int d^{15}p\ N^d{}_m(p)\ [b(p,s)u(p,s) : \exp[-iG(p, Y(y))]: + \\ + d^\dagger(p,s)v(p,s) :\exp[+iG(p, Y(y))]:\} \quad (5.125)$$

$$\psi^\dagger(Y(y)) = \sum_s \int d^{15}p\ N^d{}_m(p)\ \{b^\dagger(p,s)\bar{u}(p,s)\gamma^0 :\exp[+iG(p, Y(y))]: + \\ + d(p,s)\bar{v}(p,s)\gamma^0 :\exp[-iG(p, Y(y))]:\} \quad (5.126)$$

where : ... : denotes normal ordering and

$$G(p, Y(y)) = \int_0^{p\cdot Y(y)/m_0} m(t')dt' \quad (5.127)$$

and $N^d{}_m(p)$ a normalization constant. Contrast eqs. 5.125-5.126 to the constant mass case eqs. 5.98-5.101. The constant mass case sets $m(t') = m_0$.

If we examine the integral eq. 5.127 for a short time interval δt in the particle's rest frame then $G(p, Y(y)) \approx m(0)\delta t$ and so we define $m(0) = m_0$. Based on the formula for universe particle mass (eq. 4.8) we anticipate that m_0 might be as large as the Planck mass or larger – thus an extremely short radius. Blaha (2013) describes a quantum Big Bang model in which the initial radius of the universe is $\mathcal{O}(EM_{Planck}{}^{-2})$ where E is of the order of 1 and has the dimensions of [mass].

The spinors u(p,s) and v(p,s) are defined in a conventional way. However their form is different from the 4-dimensional case. If one takes the 256×256 $\gamma\cdot p$ matrix, then the first 128 columns give u(p,s) up to a normalization for the free particle case, the remaining 128 columns give v(p,s) up to a normalization.

Thus we have a closed form definition of a quantum universe particle wave function for universe particles of type I. A similar procedure can be followed for universe particles of types II, III, and IV.

The Feynman propagator for type I quantum fields is *not* eq. 5.109 but now has a form reflecting the Y(y) dependence of the quantum fields in eqs. 5.125 and 5.126:

$$iS_F^{TT}(y_1, y_2) = i\int \frac{d^{16}p\ \{ <0|\theta(y_{116} - y_{216})G(y_1, y_2) + \theta(y_{216} - y_{116})G(y_2, y_1)\}0>}{(2\pi)^{16}\ (p - m_0)} \tag{5.128}$$

where p^{16} is the energy and

$$G(y_1, y_2) = :\exp[-iG(p, Y(y_1))]: :\exp[+iG(p, Y(y_2))]: \tag{5.129}$$

Let

$$G_{tot}(y_1, y_2) = <0|\theta(y_{116} - y_{216})G(y_1, y_2) + \theta(y_{216} - y_{116})G(y_2, y_1)\}0> \tag{5.130}$$

$$= <0|\theta(y_{116} - y_{216}):\exp[-iG(p, Y(y_1))]::\exp[+iG(p, Y(y_2))]: +$$

$$+ \theta(y_{216} - y_{116}) :\exp[-iG(p, Y(y_2))]::\exp[+iG(p, Y(y_1)):]|0>$$

$$= <0|\theta(y^{16}{}_1 - y^{16}{}_2):\ \exp[-i\int_0^{p\cdot Y(y_1)/m_0} m(t')dt']::\exp[+i\int_0^{p\cdot Y(y_2)/m_0} m(t')dt']: +$$

$$+ \theta(y^{16}{}_2 - y^{16}{}_1):\exp[-i\ \exp[+i\int_0^{p\cdot Y(y_2)/m_0} m(t')dt']::\exp[+i\int_0^{p\cdot Y(y_1)/m_0} m(t')dt']:|0>$$

then

$$iS_F^{TT}(y_1, y_2) = i\int \frac{d^{16}p\ G_{tot}(y_1, y_2)}{(2\pi)^{16}\ (p - m_0)} \tag{5.131}$$

Except for the case of a constant mass, where $m(t) = m_0$, the Feynman propagator is not a function of $y_1 - y_2$. The evaluation of eq. 5.130 in the general case of a variable mass is straightforward but cumbersome. For the special case of a linear time dependence of the mass, $m(t) = at$, we find eq. 5.130 gives

$$G_{tot}(y_1, y_2) =<0|\theta(y^{16}{}_1 - y^{16}{}_2):\exp[-ia(p\cdot Y(y_1)/m_0)^2/2]::\exp[+ia(p\cdot Y(y_2)/m_0)^2/2]: +$$

$$+ \theta(y^{16}{}_1 - y^{16}{}_2):\exp[-ia(p\cdot Y(y_2)/m_0)^2/2]::\exp[+ia(p\cdot Y(y_1)/m_0)^2/2]:|0> \tag{5.132}$$

yielding a complex function of p, y_1, and y_2. Note that the lower bound of the integrals in the Feynman propagator cancel and thus the need for an understanding of the beginning of a universe is removed in this case.

We have shown that universe particle theory can handle the case of a variable universe mass m(t). Expanding or contracting (or oscillating) universe particles correspond to expanding and contracting (or oscillating) universes.

5.3.5 Left-Handed and Right-handed Universe Particles

In sections 5.3.1 and 5.3.2 we found that left-handed and right-handed tachyonic universe particles existed. The tachyonic nature of the universe particles indicates that their speed in the universe exceeds the "speed of light" of the Flatverse. The physical meaning of the handedness of these types of universes is an interesting issue. When we consider our universe we see left-handedness in the weak interactions of elementary particles. In addition it appears that organic molecules overwhelmingly favor left-handedness on earth although right-handed molecules exist in outer space and can be created in the laboratory. Right-handed molecules transform into left-handed molecules in watery media through electromagnetic effects.

Why nature favors left-handedness is an open question. It has given rise to speculations that gravitation, especially quantum gravitons, may be left-handed. The European Space Agency's Planck telescope will study polarization effects in the cosmos and may well be able to show that the gravitons starting from the beginning of the universe, and magnified by inflation in the universe's expansion, may be left-handed.

If handedness of gravitation is verified experimentally, then our theory of left-handed/right-handed universe particles would be substantiated. *Our universe would then be tachyonic and probably left-handed.*

5.3.6 Internal Structure of Universe Particles

We have treated universes as particles in the preceding discussion taking an extremely large view of Flatverse particles just as elementary particle theory viewed nucleons at low energies (large distances). Now we develop a detailed view of universe particles in a manner analogous to the high energy view of the internal dynamics of nucleons that led to the quark-parton model of nucleons. In the present case we shall see (chapter 6) that high energy Baryonic field probes of universe particles can yield a model of the internal baryonic structure of universes.

We know that universes are composed of matter and radiation. We believe that there is at least one interaction between universes dependent on baryon number – a baryonic, 16-dimensional gauge field. We will discuss this possibility in detail in chapter 6. In this section we will discuss the use of a baryonic gauge field to probe the baryon structure of universe particles.

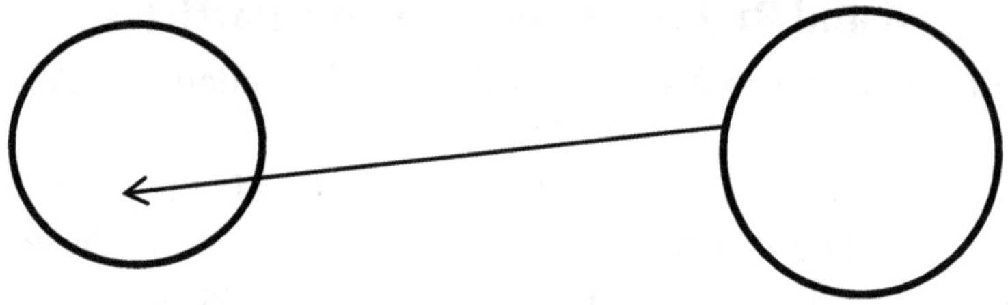

Figure 5.1. A symbolic view of a high resolution (high energy) probe from a universe to a specific baryonic part of another universe.

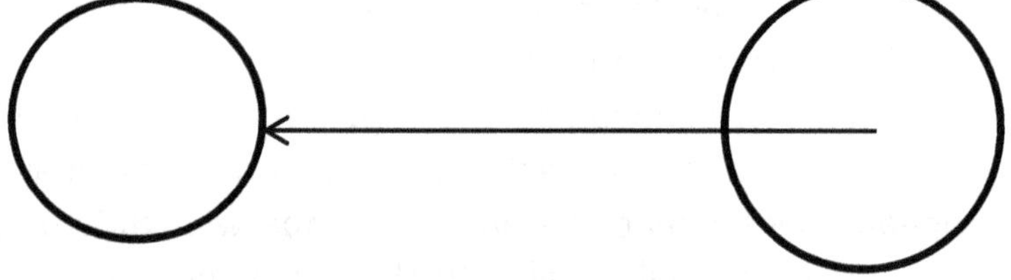

Figure 5.2. A symbolic view of a low resolution (lower energy) probe from a universe to an entire universe.

There appears to be two types of baryonic probes of a universe: 1) a series of high energy probes to specific small regions inside another universe for the purpose of mapping its internal structure; 2) a low energy probe of another universe to get a global view of its baryonic structure. The first type of probe corresponds to deep inelastic (high energy) electron-nucleon scattering which led to the quark-parton model of nucleons. The second type of probe corresponds to low energy electron-nucleon scattering to get a "global" view of a nucleon. In both case an electromagnetic (gauge) field particle (photon) was the probe particle.

Besides the inherent scientific interest in such experiments it is possible that they may be of use in the very distant future in Mankind is able to develop multiverse starships that can travel in the Flatverse to other universes. Then the baryonic gauge field becomes the "eyes" of the starship just as electromagnetic fields (light) are the eyes of current spaceships. We will consider the possibility of universe starships in a forthcoming book entitled, *All the Multiverse! Exploring the Endless Universes of Matter and Mind,* that will appear soon.

5.4 When Universes Collide: Interactions and Collisions of Universe Particles

5.4.1 Gravitation and the Fifth Force

The primary forces involved in the interactions and collisions of universe particles are the force of gravity, and a fifth force which we take to be the baryonic force described in the next chapter.[105] The force between two clumps of baryonic matter containing baryons and other particles: clump1 of mass m_1 and baryon number n_1, and clump2 of mass m_2 and baryon number n_2 is

$$F = -Gm_1m_2/r^2 + (q_B{}^2/4\pi)n_1n_2/r^2 \qquad (5.133)$$

where q_B is the baryonic constant and r is the distance between widely separated clumps. Experimentally a baryonic force between baryons has not been detected. Eötvös experiments on the ratio of the observed gravitational mass to the inertial mass showed that that it is constant to within one part in 100,000,000 as far back as 1922 indicating the baryonic force is extremely weak compared to the gravitational force. Eötvös et al[106] found

$$(q_B{}^2/4\pi)/(Gm_p{}^2) < 10^{-5}$$

where m_p is the proton mass.

---

[105] The basis for our belief is the incredible accuracy of baryon number conservation observed in experiment. This conservation law strongly suggests a gauge field symmetry for baryons and an associated force law for baryons that parallels electromagnetism.

[106] Eötvös, R. V., Pekár, D., Fekete, E., Ann. d. Physik **68**, 11 (1922).

Since then, the experiment has been redone with improved accuracy by Dicke and collaborators.[107] They have improved the accuracy to one part in 100 billion. A further analysis showed a very small discrepancy that suggested the ratio, while small, was non-zero, implying the equivalence principle might not be exact and that the discrepancy changed with the material used in the experiment – just what one might expect if a very small baryonic force was present – called the "fifth force." At present the existence and amount of the discrepancy is unclear. Nevertheless, we will assume a fifth force – baryonic force due to a baryonic gauge field that yields the baryon conservation law.

5.4.2 Universes in Collision

We assume that the dynamics of universes in collision will be analogous to that of galaxies in collision since gravity is the dominant force in both cases. Colliding galaxies have been observed often. These results will provide guidance for the case of universes in collision.[108]

It is clear in the case of colliding galaxies and of colliding large nuclei (gold and lead typically) that there are several types of collisions with differing results.

These types of universe collisions can be qualitatively classified as

1. Clean collisions in which universes nudge each other but retain their identity. These are extreme peripheral collisions. If the universes overlap slightly then the typically spherical symmetry of the universes may become distorted and they may become lopsided.[109]

2. Peripheral collisions in which the universes retain their identity but are connected by a trailing string of mass-energy. Eventually the string breaks and the universes separate. Subsequently the pieces of trailing string in each universe contract due to their universe's gravitational effects.

---

[107] P. G. Roll, R. Krotkov, R. H. Dicke, *Annals of Physics*, 26, 442, 1964.

[108] The high energy collision of atomic nuclei at Brookhaven, CERN and other laboratories also is analogous in overall detail with universes in collision.

[109] The Wilkinson Microwave Anisotropy Probe (WMAP) and the Planck European Space Agency satellite has been accumulating data since 2001 that suggests the universe may be lopsided with hot and cold spots on opposite sides of the universe differing from those on the other side being hotter and colder respectively. Perhaps the result of a collision when the universe was young.

3. Two universes can collide and produce multiple universes.

4. Two universes can collide in a "central" collision and amalgamate into one universe.

We will discuss universe interactions in more detail in chapter 6.

5.5 Bosonic Universe Particles

The previous section has described fermionic universe particles. In this section we will briefly describe aspects of bosonic (spin 0) universe particles. First it is important to note that the Wheeler-DeWitt equation being second order like the Klein-Gordon equation seems to suggest that universe particles should be bosonic – like Klein-Gordon equation particles.[110] The Wheeler-DeWitt equation has a mass-like term R that can be positive or negative. If the mass term is negative then the wave-like propagation of the state functional (wave equation solution) can be in space-like directions implying a tachyonic solution. Thus the Wheeler-DeWitt equation supports "normal" state functionals that propagate in time-like directions as well as tachyonic propagation.

Partly for this reason we suggest that bosonic universe particles can be either normal or tachyonic. Tachyonic bosonic universe particles can fission in a manner similar to tachyonic fermionic universe particles. The fission equations of section 5.5.2 also apply to tachyonic bosonic universe particles.

The quantum field theory of normal and tachyonic bosonic universe particles is similar to that of ordinary bosons. See Blaha (2005a) for the boson case discussion that is paralleled by our universe particle formalism.[111]

5.6 Physical Meaning of Universe Particle Spin

The physical meaning of spin is a continuing discussion topic. We have suggested that spin states are in essence logic states with changes in spin an analogue to changes in logical values in a discourse or computer program. Since

---

[110] One should remember that the Wheeler-DeWitt equation is not in space but in a 6-dimensional manifold, denoted M, of metrics with one "time" dimension – having hyperbolic signature – + + + + + when the metric is positive definite. See DeWitt's paper.

[111] We leave that development as an exercise for the interested reader.

the matrix formalism for spin ½ and higher spin states is formally similar to the formalism for angular momentum, one can combine spin and angular momentum as we do in quantum theory.

In the case of universe particles, one can also associate universe particles with "true" and "false" values. Fermionic universe states have 128 truth values and correspond to a multi-valued logic. The numerousness of truth values is due to the 16-dimensional space within which universe particles reside.

Naturally one would like to know the physical differences between these 128 types of universe particles. Does the difference reside in different shapes of the universe particles? Or is the difference somehow a consequence of the global mass-energy distribution of the universe that we have not been able to discern since we only know of one universe?

The physical meaning of spin for elementary particles is also somewhat elusive. It does not reflect the flow of charge within a particle. For if it did reflect physical spinning of a particle the outer edges of a particle such as an electron would be traveling at a speed faster than light. So spin is not a mechanical property of the internal structure of an elementary particle. We have suggested that it is a truth value in the matrix formulation of a 4-valued logic called Asynchronous Logic. Thus it has no certain tangible physical basis.

In the case of universe particles the situation is unclear at present. It could be taken to be a reflection of the structure of mass-energy within a universe. This view would be contrary to our proposed view of elementary particle spin as truth values. So we can only assert that a logic interpretation is the only sensible one (based on our present knowledge or our lack thereof). The physical role of universe particle spin is only evident in interactions between universe particles via the baryonic gauge field proposed in the next chapter. Thus one must simply view it as a construct for the present.

Other than our mapping of spin values to logic values in Asynchronous Logic there is little anyone can say about the physical origin of elementary particle spin. Specifying a symmetry group as the origin, such as SuperSymmetry, is not sufficient.

5.7 Elementary Particles with Time Dependent Masses

The discussions of this chapter were presented for universe particles. However they could also apply to elementary particles or condensed matter excitations with some changes – primarily changes due to different dimensions.

Elementary particles with time dependent masses do not exist as far as we know. And that is a good thing. The idea of an elementary particle expanding indefinitely, like a universe, would have disastrous consequences for life as would particles contracting indefinitely. However, particles with oscillating masses would seem to be physically acceptable. The development of an elementary theory with oscillating masses would be an interesting exercise that might have applications in condensed matter physics.

5.8 Impact of Universe Particle Acceleration - Lopsided Internal Structure of Universe

We are developing a theory of universe particle interactions. Such interactions would cause universe particles to accelerate and should be detectable within the universe as a "lopsidedness" – there would be a shift of parts of the universe away from the direction of acceleration resulting in a difference in the features of the universe "in front" compared to those "in back" – an acceleration effect just as one sees when a jet accelerates.

Interestingly new data from the Planck observatory of the European Space Agency confirms and extends earlier data from NASA's WMAP observatory that one side of the universe appears different from the other side. There are temperature differences and mass distribution differences – just as one might expect if the universe were accelerating as a unit.

Thus we see the beginning of data suggesting our universe may be moving – "indeed accelerating" – through a multiverse. Some Planck observatory scientists have suggested their data is a preliminary indication of the multiverse.

6. Baryonic Gauge Field, Baryon Conservation, Quantum Observability, Impact on Gravitation, and Universes from Baryonic Gauge Field Vacuum Fluctuations

6.1 Flatverse Baryonic Gauge Field - Plancktons

The conservation of baryon number has been repeatedly investigated by experimenters and found to be true to extremely high accuracy. For decades theorists have suggested that the conservation law follows from the existence of an abelian gauge field in a manner much like electric charge conservation follows from the properties of the electromagnetic abelian gauge field.[112]

We will therefore assume an abelian baryonic gauge field exists that is similar to the electromagnetic field except for features due to its definition and existence in the 16-dimensional Flatverse. This field will couple extremely weakly to individual baryons as well as universe particles with non-zero baryon number.[113] We will call the baryonic gauge field particle a *planckton*. Its electromagnetic analogue is the photon.

Plancktons propagate in the Flatverse, both within universes, and exterior to universes. So the planckton field will be defined in 16-dimensional Flatverse coordinates. They will interact with baryons within a universe with Flatverse coordinates mapped to the curved coordinates in the universe. (This mapping was discussed earlier in detail.)

Since a planckton field in 16-dimensional conventional coordinates would lead to divergences we will use quantum coordinates:

$$Y^i(y) = y^i + i\, Y_u{}^i(y)/M_u{}^8 \qquad (5.4)$$

---

[112] See Gell-Mann, M. and Levy, M. *Nuovo Cimento* 16, 705 (1960) for a proof.

[113] See section 5.4.1 for a discussion of the strength of its interactions relative to gravitation.

with quantum coordinate derivatives defined by

$$\partial_i = \partial/\partial Y^i(y) = \partial/\partial(y^i - Y_u^{\,i}(y)/M_u^{\,8}) \qquad (5.5)$$

to obtain a completely finite theory of planckton interactions with elementary particles and universe particles.

Plancktons and the $Y_u^{\,i}(y)$ field of quantum coordinates are the only fields in the space between universes in the Flatverse. Since the mass-energy and charge of universes is zero, gravitation and Standard Model fields are zero in the space between universes.[114]

6.2 Beyond the Planckton

The analogy between plancktons and the electromagnetic field photons raises the possibility that the baryonic gauge field may be one of a set of gauge fields embodying an SU(2)⊗U(1) symmetry like ElectroWeak theory. We will not pursue this possibility since it is extremely unlikely to be testable in the foreseeable future.

The possibility of more complex symmetries is supported by the complex 16-dimensional Flatverse space and the need for a Reality group to make Flatverse space and time coordinates real-valued, just as in the case of the complex 4-dimensional space-time of our universe which led to the Reality group for our universe SU(3)⊗SU(2)⊗U(1)⊗SU(2)⊗U(1) and further led to The New Standard Model.

6.3 Planckton Second Quantization

The second quantization of the free planckton field $B_u^{\,i}(y)$ is similar to the second quantization of the quantum part of the Flatverse quantum coordinates $Y_u^{\,i}(y)$ as defined in section 5.3.3. The purpose and role of these fields is quite different: the planckton field generates an interaction between baryons while the $Y_u^{\,i}(y)$ field serves as the quantum part of 4-dimensional quantum coordinates giving us a finite quantum field theory of The New Standard Model and gravitation as well as a finite Big Bang for our universe.

---

[114] The vacuum energy of the baryonic field and the $Y_u^{\,i}(y)$ fields being uniform throughout the Flatverse do not exert forces or cause gravitational effects except possibly through baryonic Casimir forces between universes.

We begin by noting that Flatverse quantum coordinates are defined by eqs. 5.4 and 5.5 above. The lagrangian density terms for the free $B_u^{\,i}(Y(y))$ fields is

$$\mathscr{L}_{Bu} = -\tfrac{1}{4}\, F_{Bu}{}^{\mu\nu}(Y(y))F_{Bu\mu\nu}(Y(y)) \tag{6.1}$$

with Y(y) given by eq. 5.4 and the lagrangian is

$$L_{Bu} = \int d^{15}y\, \mathscr{L}_{Bu}(Y(y)) \tag{6.2}$$

with

$$F_{Bu\mu\nu} = \partial B_{u\mu}(Y(y))/\partial Y^{\nu}(y) - \partial B_{u\nu}(Y(y))/\partial Y^{\mu}(y) \tag{6.3}$$

where the values of μ and ν range from 1 to 16 in this section.

The equal time commutation relations, derived in the usual way, are:

$$[B_u{}^{\mu}(Y(\mathbf{y}, y^0)), B_u{}^{\nu}(Y(\mathbf{y}', y^0))] = [\pi_u{}^{\mu}(Y(\mathbf{y}, y^0)), \pi_u{}^{\nu}(Y(\mathbf{y}', y^0))] = 0 \tag{6.4}$$

$$[\pi_{uj}(Y(\mathbf{y}, y^0)), B_{uk}(Y(\mathbf{y}', y^0))] = -i\,\delta^{15tr}{}_{jk}(Y(\mathbf{y},0) - Y(\mathbf{y}',0)) \tag{6.5}$$

where

$$\pi_u{}^{k} = \partial \mathscr{L}_u(B_u(Y(y)))/\partial B_{uk}'(Y(y)) \tag{6.6}$$

$$\pi_u{}^{0} = 0 \tag{6.7}$$

and

$$\delta^{tr}{}_{jk}(\mathbf{y} - \mathbf{y}') = \int d^{15}k\, e^{i\,\mathbf{k}\bullet(Y(\mathbf{y},0) - Y(\mathbf{y}',0))}\,(\delta_{jk} - k_jk_k/\mathbf{k}^2)/(2\pi)^{15} \tag{6.8}$$

$$B_{uk}'(Y(y)) = \partial B_{uk}(Y(y))/\partial y^{16} \tag{6.9}$$

for j, k = 1, 2, … , 15.

If we choose the Coulomb gauge for $B_{uk}(Y(y))$:

$$B_u{}^{16}(Y(y)) = 0$$

$$\partial B_u{}^{j}(Y(y))/\partial Y^{j}(y) = 0$$

for j = 1, 2, … , 15 then fourteen degrees of freedom are present in the vector potential.[115] The Fourier expansion of the vector potential $B_u^{\,i}(Y(y))$ is:

---

[115] Note we use the Coulomb gauge for Y(y) also.

$$B_u^{\,i}(Y(y)) = \int d^{15}k\ N_{0B}(k) \sum_{\lambda=1}^{14} \varepsilon^i(k,\lambda)[a_B(k,\lambda) :e^{-ik\cdot Y(y)}: + a_B^{\dagger}(k,\lambda) :e^{ik\cdot Y(y)}:] \quad (6.10)$$

for i = 1, … , 15 where

$$N_{0B}(k) = [(2\pi)^{15} 2\omega_k]^{-1/2} \quad (6.11)$$

and (since the field is massless)

$$k^{16} = \omega_k = (\mathbf{k}^2)^{1/2} \quad (6.12)$$

where k^{16} is the energy, and where the $\varepsilon^{\,i}(k, \lambda)$ are the polarization unit vectors for $\lambda = 1, \ldots, 14$ and $k^\mu k_\mu = k^{16\,2} - \mathbf{k}^2 = 0$.

The commutation relations of the Fourier coefficient operators are:

$$[a_B(k,\lambda), a_B^{\dagger}(k',\lambda')] = \delta_{\lambda\lambda'}\delta^{15}(\mathbf{k} - \mathbf{k'}) \quad (6.13)$$
$$[a_B^{\dagger}(k,\lambda), a_B^{\dagger}(k',\lambda')] = [a_B(k,\lambda), a_B(k',\lambda')] = 0 \quad (6.14)$$

and the polarization vectors satisfy

$$\sum_{\lambda=1}^{14} \varepsilon_i(k,\lambda)\varepsilon_j(k,\lambda) = (\delta_{ij} - k_i k_j/\mathbf{k}^2) \quad (6.15)$$

The $B_u^{\,\mu}$ Feynman propagator is

$$iD_F^{\,trTT}(y_1 - y_2)_{jk} = \langle 0|T(B_{uj}(Y(y_1))B_{uk}(Y(y_2)))|0\rangle \quad (6.16)$$

$$= -\,ig_{jk} \int \frac{d^{16}k\ e^{-ik\cdot(y_1 - y_2)}\ R(\mathbf{k}, y_1 - y_2)}{(2\pi)^{16}\ (k^2 + i\varepsilon)} \quad (6.17)$$

where g_{jk} is the 16-dimensional Lorentz metric and where $R(\mathbf{k}, y_1 - y_2)$ is given by

$$R(\mathbf{k}, y_1 - y_2) = \exp[-\,k^i k^j \Delta_{Tij}(y_1 - y_2)/M_u^{\,16}] \quad (6.18)$$
$$= \exp\{-k^2[A(v) + B(v)\cos^2\theta] / [(2\pi)^{14} M_u^{\,4} z^2]\}$$

with

$$z^\mu = y_1^{\,\mu} - y_2^{\,\mu}$$
$$z = |\mathbf{z}| = |\mathbf{y_1} - \mathbf{y_2}|$$
$$k = |\mathbf{k}|$$

$$v = |z^0|/z$$
$$A(v) = (1 - v^2)^{-1} + .5v \ln[(v - 1)/(v + 1)]$$
$$B(v) = v^2(1 - v^2)^{-1} - 1.5v \ln[(v - 1)/(v + 1)]$$
$$\mathbf{k{\cdot}z} = kz \cos\theta$$

and $|\mathbf{k}|$ denoting the length of a spatial 15-vector **k** while $|z^0|$ is the absolute value of $z^0 \equiv z^{16}$.

As eq. 6.18 indicates, the Gaussian damping factor R(k, z) for all large spatial momentum k^j is the same for both the positive and negative frequency parts of the (two-tier) B_u Feynman propagator. We are assuming the spatial momentum is real-valued in this discussion. It is also important to note that R(k, z) does not depend on $k^0 = k^{16}$ (in the B_u and Y_u Coulomb gauges) and thus the integration over k^0 proceeds in the usual way to produce time-ordered positive and negative frequency parts.

The Gaussian exponential factor in eqs. 6.17 – 6.18 causes the Feynman propagator to be finite and, together with the Gaussian factor in universe particle propagators, causes all perturbation theory calculations when interactions are introduced to be finite as we have seen earlier in The New Standard Model.

For small momentum much less than M_u then $R(\mathbf{k}, y_1 - y_2) \to 1$ and the Feynman propagator is the "normal" propagator of conventional 16-dimensional quantum field theory. For large momentum the corresponding potential approaches r^{13} in contrast to the electromagnetic Coulomb potential r^{-1}. The B_u potential is highly non-singular at large energies.

6.4 Planckton Interactions with Universe Particles and Individual Baryons

In this section we will develop an interacting theory of universe particles and plancktons from the lagrangian terms of universe particles (eq. 5.59), plancktons (eq. 6.1) and quantum coordinates (eq. 5.70). We will only consider the case of type I universe particles since the other cases differ from it only in details.

$$\mathcal{L} = \bar{\psi}(Y(y))[i\gamma^\mu\partial/\partial y^\mu - e_B\gamma^\mu B_{u\mu}(Y(y)) - m(t)]\psi(Y(y)) - \tfrac{1}{4} F_{Bu}{}^{\mu\nu}(Y(y))F_{Bu\mu\nu}(Y(y)) - - \tfrac{1}{4} F_u{}^{\mu\nu}(y)F_{u\mu\nu}(y) \qquad (6.19)$$

where μ, ν = 1, 2, … , 16 and where

$$\bar{\psi} = \psi^{\dagger}\gamma^{16}$$

$$F_{Bu_{\mu\nu}} = \partial B_{u_\mu}(Y(y))/\partial Y^{\nu}(y) - \partial B_{u_\nu}(Y(y))/\partial Y^{\mu}(y) \quad (6.3)$$

$$F_{u_{\mu\nu}} = \partial Y_{\mu}/\partial y^{\nu} - \partial Y_{\nu}/\partial y^{\mu} \quad (5.71b)$$

$$Y^{i}(y) = y^{i} + i\, Y_{u}^{i}(y)/M_{u}^{8} \quad (5.4)$$

$$e_B = e_{B0}/M_u^{6} \quad (6.20)$$

with e_{B0} a dimensionless coupling constant, and with μ and ν ranging from 1 through 16.

The lagrangian is

$$L = \int d^{15}y\,\mathscr{L} \quad (6.21)$$

Note the dimensions of the fields differ in the 16 dimensional space:

$$Y^{\mu} \sim [\text{mass}]^{7}$$

$$B_{u_\mu} \sim [\text{mass}]^{7}$$

$$\psi \sim [\text{mass}]^{15/2}$$

as can be seen from the above lagrangian as well as earlier equations. Note also that the mass and thus the size of universe particles is time dependent. They can expand or contract with time depending on their internal characteristics (gravitation and effects of elementary particle interactions) which are not embodied in this lagrangian. As a result this theory, incomplete as it is, does not conserve energy unless m(t) is constant.

The lagrangian generates the baryonic interactions of universe particles using Two Tier quantum coordinates which prevent infinities in perturbation theory calculations.

The interaction of baryon elementary particles with the baryonic field requires terms in The New Standard Model specifying the baryon field interaction baryons with the form

$$e_B\gamma^{\mu}B_{u_\mu}(Y(y))$$

The following sections describe some of the physically significant interactions that the lagrangian (eq. 6.19) implies.

6.5 Creation of Universes through Baryonic Gauge Field Fluctuations

One of the most exciting questions in Cosmology is the origin of our universe. The conventional view is that it originated in a Big Bang from an infinitesimal point in space. The source of the Big Bang and the prior state of the Cosmos, if there was one, is the subject of much speculation. Based on the particle interpretation of the Wheeler-DeWitt equation, the possibility of a baryonic force strongly supported by conservation of baryon number, and the multiverse concept it is reasonable to consider the possibility that the universe originated in a vacuum fluctuation.

Our formulation of universe particle theory provides for the generation of a universe particle and anti-particle as a vacuum fluctuation. We view a universe particle as having a substantial excess of baryons, N, as we see in our universe. Its anti-universe at the time of creation (the Big Bang point) is its "mirror image" having the same number of anti-baryons (baryon number –N) so that baryon number is conserved bý the fluctuation event.

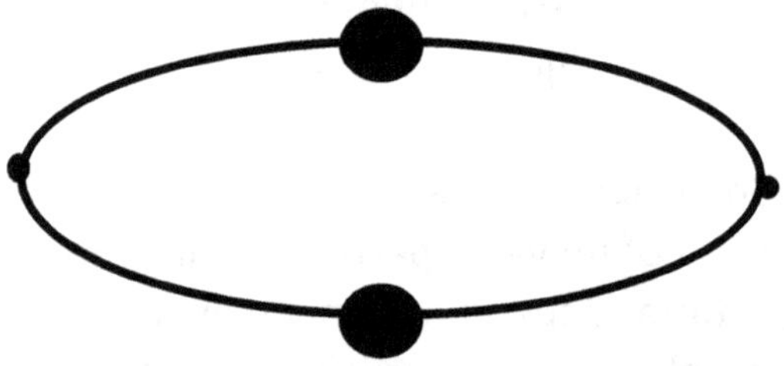

Figure 6.1. Generation of a universe – anti-universe pair as a vacuum fluctuation.

The small value of the coupling constant should lead to an extremely long lifetime for the universes generated by the fluctuation. Thus the 13.7 billion year life of our universe is not unreasonable. Its lifetime should be extremely long. The probability of the creation of universes by vacuum fluctuations should be correspondingly small.

6.6 When Universes Collide: Coalescence of Universes

Universes moving in the Flatverse can collide through chance, or due to the planckton field which causes universes with excess baryons to attract universes with excess anti-baryons.

When universes collide several possibilities present themselves:

1. They can graze each other distorting each other's shape and internal baryon distribution through the baryonic force while maintain their individual identity.

2. They can intermix with both the baryonic and gravitational forces causing a redistribution of their masses. They may separate afterwards or may coalesce into a single universe. One result of this may be lopsided universes. Our universe appears to be lopsided. Some cosmologists believe this is due to a near collision of our universe with another shortly after the Big Bang.

6.7 Fission of Universes

Under certain circumstances the distribution of matter in the universe may lead to the fission of the universe into two separate universes. Our model lagrangian supports this possibility for universe particles. The detailed mechanism of the fission process is not specified by the model.

6.7.1 Fission of Normal universes

The fission of universe particles in our universe particle model is depicted in the Feynman diagram in Fig. 6.2.

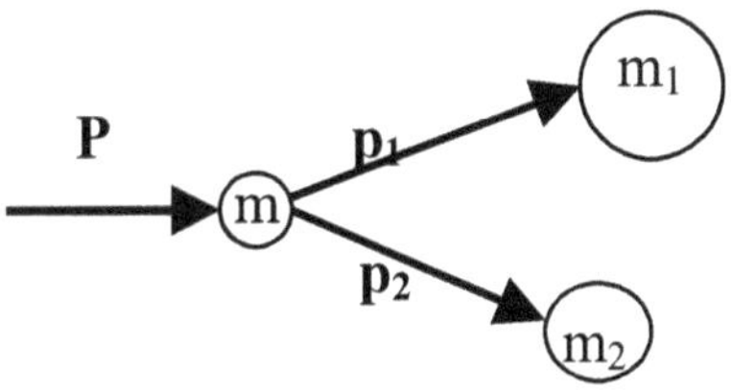

Figure 6.2. Fission of a universe particle into two universe particles.

The sum of the masses of the output universe particles is usually less than the original universe particle mass. However if the fission takes a long time and the masses are time dependent then the produced universe particles combined masses may exceed the original universe's mass.

6.7.2 Tachyon Universe Particle Fission to More Massive Universe Particles

In Blaha (2007a) we showed that a tachyonic (faster than light) particle could fission into particles of larger mass. In this section we will show that a tachyonic universe particle may fission into two more massive universe particles. This phenomenon is of particular interest because it enables tachyonic universes to spawn in a new novel way not previously considered in discussions of the origin of universes.

The lagrangian for a tachyonic universe particle is

$$\mathscr{L}_{||} = \psi_T^{\ S}(Y(y))[\gamma^\mu\partial/\partial y^\mu - e_B\gamma^\mu B_{u_\mu}(Y(y)) - m(t)]\psi(Y(y)) - \tfrac{1}{4} F_{Bu}{}^{\mu\nu}(Y(y))F_{Bu_{\mu\nu}}(Y(y)) - \\ - \tfrac{1}{4} F_u{}^{\mu\nu}(y)F_{u_{\mu\nu}}(y) \qquad (6.22)$$

based on eqs. 6.19 and 5.60 and 5.61. We assume m(t) is constant.

When a particle or a universe particle fissions (decays) one normally expects that the masses of the particles or universe particles produced by the decay to be smaller than the mass of the original particle or nucleus. In the case of tachyonic (faster-than-light) elementary particles or universe particles a much different possibility is present: a tachyon can decay into heavier tachyons. We will consider the specific case of a tachyon universe particle decaying into two universe particles whose total mass is greater than the original. (See Fig. 6.3.)

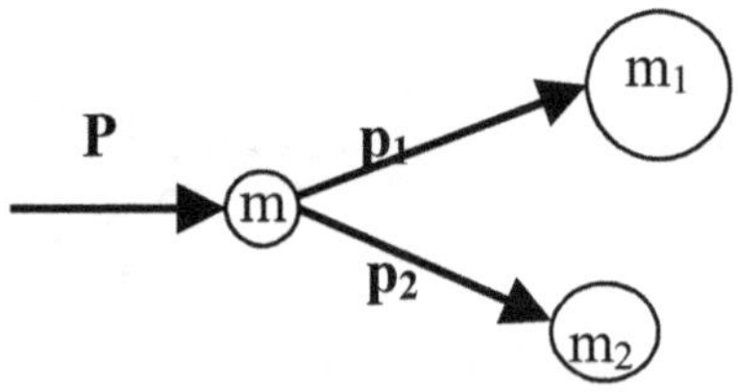

Figure 6.3. Two universe particle decay of a tachyon universe particle.

We will assume the initial tachyon universe particle has zero energy ($p^{16} = 0$) and thus the tachyons universe particles emerging from the decay also have total universe particle energy zero. The analysis is based on conservation of total universe energy and momentum. The below discussion applies to 16-dimensional space with 15-dimensional spatial coordinates.

Momentum conservation implies

$$\mathbf{P} = \mathbf{p}_1 + \mathbf{p}_2 \tag{6.23}$$

Since all energies are zero

$$\begin{aligned}(cP)^2 &= (c\mathbf{P})^2 = m^2 \\ (cp_1)^2 &= (c\mathbf{p}_1)^2 = m_1^2 \\ (cp_2)^2 &= (c\mathbf{p}_2)^2 = m_2^2\end{aligned} \tag{6.24}$$

where $P = |\mathbf{P}|$, $p_1 = |\mathbf{p}_1|$, and $p_2 = |\mathbf{p}_2|$. If we now square eq. 6.23 and then use eqs. 6.24 we obtain

$$m^2 = m_1^2 + m_2^2 + 2m_1m_2 \cos\theta \tag{6.25}$$

where θ is the angle between the emerging universe particles momenta $\mathbf{p}_1$ and $\mathbf{p}_2$. Eq. 6.25 has a number of interesting cases:

Case $\theta = 0$:

$$m = m_1 + m_2 \tag{6.26}$$

The masses of the outgoing universe particles sum to the mass of the original tachyon universe particle.

Case $\theta = \pi/2$:

$$m^2 = m_1^2 + m_2^2 \tag{6.27}$$

The masses of each outgoing universe particle tachyon is less than the mass of the original tachyon universe particle.

Case $\theta = \pi$:

$$m^2 = (m_1 - m_2)^2 \tag{6.28}$$

In this case either $m_1 > m$ or $m_2 > m$. Thus one of the outgoing tachyon universe particles has a greater mass than the original tachyon universe particle. Mass is effectively created from the spatial momentum of the initial universe particle. This process is the inverse of normal particle and universe particle fission where the sum of the outgoing masses is always less than the original particle's mass and the difference is mass converted into energy in the form of additional photons.

This last case, where one of the outgoing universe particles is more massive than the original universe particle, is not just for $\theta = \pi$. Since

$$\cos\theta = (m^2 - m_1^2 - m_2^2)/(2m_1m_2) \tag{6.29}$$

we see that the sum of the outgoing universe particle masses is always greater than the original tachyon universe particle *mass (except when $\theta = 0$)* since

$$\cos\theta = 1 + [m^2 - (m_1 + m_2)^2]/(2m_1m_2) \leq 1 \tag{6.30}$$

and thus

$$[m^2 - (m_1 + m_2)^2]/(2m_1m_2) \leq 0 \tag{6.31}$$

Note $m = m_1 + m_2$ only if $\theta = 0$.

Since we can transform the above discussion to the case of universe particle tachyons having non-zero energy using an ordinary 16-dimensional Lorentz transformation the discussion in this subsection is general.

We therefore conclude that when a tachyon universe particle decays into two tachyon universe particles the sum of the masses of the produced tachyon universe particles is greater than the mass of the original tachyon universe particle except if the angle between the momenta of the produced tachyon universe particles is zero. In that case the sum of the masses of the produced tachyon equals the mass of the original tachyon universe particle and the produced universe particles overlap.

6.8 Universe Particle – Planckton Interactions

These interactions are quite similar to two-tier electromagnetic interactions except that universe particles have time-dependent masses, and that the space is 16-dimensional.

The interactions have a new aspect due to the time dependence of the universe particle masses. This feature is illustrated by Fig. 6.4: the mass of a universe particle after a baryonic interaction vertex is the same as it was before the interaction assuming the point-like interaction specified in the lagrangian.

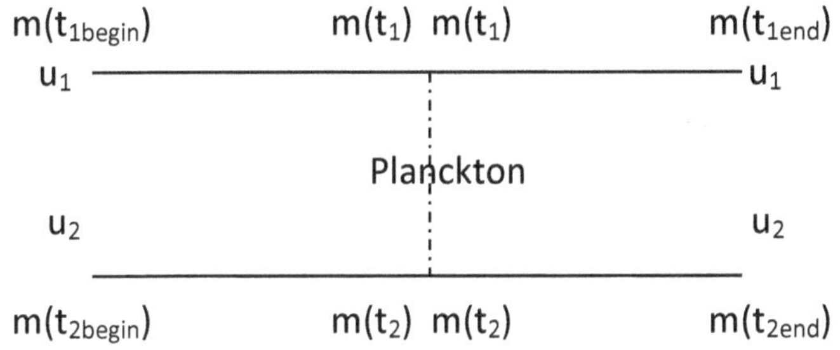

Figure 6.4. A Feynman diagram illustrating the continuity of a universe particle mass through a Planckton interaction.

The reader may verify this by writing the perturbation theory equivalent. A universe particle vertex corresponds to

$$iS_F^{TT}(y_1, y_2)\gamma^\mu iS_F^{TT}(y_2, y_3) \tag{6.32}$$

By eqs. 5.131 and 5.132 the universe particle mass is the same on either side of the interaction vertex.

6.9 Internal Structure of Universe Particles

We have developed a planckton field theory that gives interactions between baryons. This theory is applicable to universe-universe interactions. It also yields baryon particle – baryon particle interactions as well as baryon particle – universe particle interactions.

It is possible for a planckton to be emitted in one universe and interact with a baryon elementary particle in another universe. This type of "probe" must be a high energy probe just as a photon probe of the internal structure of a nucleon[116] must be a high energy photon to bring out the nucleon's internal structure (parton model).

In this section we will discuss planckton probes of other universes, and the internal structure of a universe as a mass distribution governed by gravitation as it relates to universe particles.

6.9.1 Planckton Probes

Plancktons can be generated in one universe and be used to probe the baryon distribution of another universe. Since the planckton propagator (eq. 6.16) is expressed in Flatverse coordinates the baryon distribution in the target universe will be a distribution in Flatverse coordinates. Flatverse coordinates can be expressed in terms of the curved space-time coordinates of a universe x^μ using eq. 1.4. However the inversion of eq. 1.4

$$x^\mu = f^{-1\mu}(y) \tag{6.33}$$

is not 1:1 since x^μ is 4-dimensional and y is a 16-dimensional vector. The universe coordinates x^μ are each individually determined up to a subspace. One might be concerned about this situation but the determination of the distribution in Flatverse coordinates gives a more direct picture not convoluted by the curvature of the target universe.

The detailed probing of a target universe requires high energy plancktons. The similarity of this procedure to deep inelastic electro-nucleon scattering is obvious to the high energy physicist. But in doing this planckton probe experiment one obtains a picture of a different universe – something that is not possible to do with electromagnetic or graviton probes.

6.9.2 Internal Structure of a Universe Particle

The development of the theory of universe particles which resulted in the lagrangian of eq. 6.19 does not fully describe universe particles since it neglects the internal structure of a universe particle. The internal structure of a universe particle is primarily determined by gravitation, electromagnetic effects and nuclear physics.

---

[116] Deep inelastic electron-nucleon scattering.

Consequently the full lagrangian of a universe particle has the form

$$\mathcal{L}_{tot} = \mathcal{L}_{internal} + \mathcal{L} \tag{6.34}$$

where $\mathcal{L}$ is determined by eq. 6.19. As a result the complete quantum wave function of a universe particle has the form

$$\psi_{tot} = \psi_{internal}(Y)\psi_{ext}(Y) \tag{6.35}$$

where $\psi_{internal}(Y)$ is the internal wave function and $\psi_{ext}(Y)$ is determined by eq. 6.19. It seems reasonable to have a separable equation except when universes collide. In that situation a perturbative mixing of the universes and their wave functions applies and it may be possible to calculate the collision output universes by introducing a further interaction between the internal and external aspects of the universe particles.

6.10 Conservation of Baryon Number

The conservation law for baryon number has been derived in many articles and books. See Gell-Mann et al[117] and Sakurai (1964) for excellent discussions.

[117] Gell-Mann, M. and Levy, M. *Nuovo Cimento* 16, 705 (1960).

7. The Flatverse as a New Level of Physical Reality

7.1 A New Level of Physical Phenomena – a Flatverse of Universes

With the development of the multiverse/Flatverse concept we add a new level to the natural hierarchy of creation: universes in the Flatverse. While the existence of a Flatverse filled with many universes is highly speculative, the existence of a sister universe, which is well-motivated by earlier discussions, suggests that having additional universes is reasonable.[118]

These universes may be so numerous that they may constitute a dilute gas of universes. The distance scales in our universe range from light years to billions of light years. The distance scales in the Flatverse range from billions of light years to trillions of light years or more. (The Flatverse may well have an infinite number of universes dispersed within it extending to infinite distance.)

Presently, and for the foreseeable future, the multiverse speculation cannot be proven – or disproven. If universes collide then the possibility exists, although it is very remote, that quantum entanglement might lead to evidence for the existence of other universes.

7.2 Features of the Multiverse

There are a number of features of the multiverse/Flatverse that follow from previous discussions which we may believe to be true:

- The Flatverse space is flat – there is no gravitation between universes. (Otherwise the Flatverse would not be flat.)

- There is a baryonic abelian gauge field force between universes. This force can also exist between a universe and the baryons within the same universe, and between a universe and the baryons within another universe. The phenomenon is analogous to electromagnetic electron-proton scattering and proton-proton scattering. At low energies the force is between universes. At high energies baryons in different universes interact directly.

---

[118] Based on the maxim: "What is not forbidden is allowed."

- The "mass" of a universe is time dependent and proportional to the area of the universe in a manner analogous to the case of black holes.

- Universe particles may have a spinor formulation.

- Each universe occupies a region of the Flatverse that is a closed set in Flatverse coordinates and an open set in the curved coordinates of the universe.

- Within the boundaries of a universe one can use either Flatverse coordinates or the curved coordinates of the universe.

- The Baryon gauge field can create universe-antiuniverse pairs through vacuum fluctuations.

- Due to their size the creation or interaction of universes via the baryonic field will have form factors or structure constants analogous to similar features in the electromagnetic behavior of hadrons.

- It is possible that universes can be tachyonic. This is explicitly seen in the solutions of the Wheeler-DeWitt equation.

- It is possible that universes also have additional, ultra-weak interactions. Perhaps they have an SU(2)⊗U(1) symmetry analogous to the known ElectroWeak interaction. Perhaps they have a full set of Standard Model-like interactions: a universes' Reality group consisting of SU(3)⊗SU(2)⊗U(1)⊗SU(2)⊗U(1), may apply to universes if equivalents to color and other internal quantum numbers exist to label universes. These "internal" quantum numbers may well be a global feature of a universe and thus not detectable by experiments within the universe. Since the Flatverse is complex the appearance of a Flatverse Reality group is well motivated.

8. Prospectus for the Future of Fundamental Physics and Cosmology

In a universe the fundamental particles and dynamics are based on Asynchronous Logic, complex space-time and a map from complex space-time to the real physical space-time of our experience. The existence of universes within the Flatverse gives a preferred status to inertial reference frames. In particular, the existence of a separate sister universe provides mass, inertia, a quantum observer and a time scale for its sibling universe. These features seem to precisely specify the known fundamental features of our universe.

8.1 Is There a More Fundamental Layer?

The generality of the features strongly suggests that they may apply to all universes which may also be paired.

One may ask whether there is yet a more fundamental layer of physics which underpins the above scenario. Given the simplicity of the fundamental assumptions it is difficult to believe that known physics is a subsector of a larger physics.

The only honest answer to this question is further experiment to find new physical features beyond those in our extended Standard Model and Cosmology. It is possible that we have only found a subset of physics and more will be found. Only more experimentation will tell. So like Prufrock[119] (and Hesiod[120]) we must continue to "spit out all the butt ends[121] of our days and ways" promoting physical research while realizing that we are only at the very beginning in science and civilization. Twelve thousand civilizing years is a very short time.

---

[119] T. S. Eliot, *The Love Song of J. Alfred Prufrock*. In this line Eliot alludes to Hesiod's *Works and Days* (Hesiod's poem of advice to his brother. c. 600 BC) – part of the universality extent in Prufrock. This confluence of thought over a range of 2600 years illustrates this author's view that we are only at the beginning of human advancement: not just in science but in all fields of human endeavor. I heard Prufrock's lines spoken by Mr. Eliot at the NY YMHA in 1961 in his unforgettable voice.

[120] Hesiod was a farmer-poet who rivaled Homer in classical times. There is a famous poem, "The Contest of Homer and Hesiod" describing a contest between Hesiod and Homer which Hesiod won. The judge, King Paneides of the Chalcidians, gave the crown of victory to Hesiod (although Homer was the better poet) saying "it was right that he who called upon men to follow peace and husbandry should have the prize rather than one who dwelt on war and slaughter." p. 587 Evelyn-White (1914). A judgment showing Mankind's long yearning for peace as we enter more war-like times.

[121] Cigarette butts.

Appendix. Some Features of Complex General Relativity

The following chapters are reproduced from Blaha (2004), as is, to show some novel features of complex General Relativity..

Appendix 3A. Cauchy Theorem, Parallel Transport, Gauss' Theorem, Volume and Area Integrals, a Four Index Symmetric Curvature, and Complex Determinants in Complex 4-Dimensional Space-Time

3.7.5 Line Integral in Several Complex Coordinates around a Closed Curve – Generalized Cauchy Theorem

In the theory of one complex variable, Cauchy's theorem states the line integral of a function around a closed curve C is zero if the function is holomorphic on C and within the region enclosed by C. We now ask whether Cauchy's theorem can be extended to complex space-time.

Consider a closed curve C in complex space-time. The curve C is assumed to follow a path that involves several complex dimensions (see Fig. 3.7.5.1).

Figure 3.7.5.1 A symbolic depiction of a closed curve in complex space-time.

A line integral around C can be viewed as a sum of line integrals around rectangular patches in a sufficiently fine grid (in the limit that the rectangles in the grid become infinitesimal in size.) See Fig. 3.7.5.2. The line integral contributions from the common interior lines cancel in pairs so that the sum of the line integral contributions of the infinitesimal rectangles is equal to the sum of the uncanceled contributions from the outer edges along C.

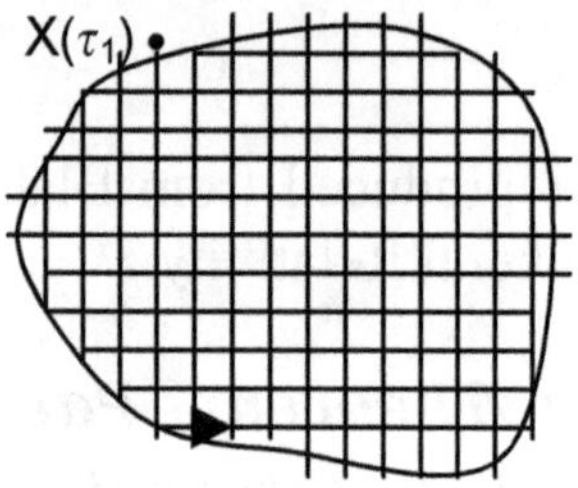

Figure 3.7.5.2 A closed curve in complex space-time. A grid of sufficiently fine mesh is superimposed on the area enclosed by the curve.

Thus the question of whether the integral of a holomorphic function around a closed space-time curve is zero or not can be reduced to a consideration of the line integral around an infinitesimal rectangle. To that end we will consider the line integral

$$I_c = \oint_c Z_\mu(x)dx^\mu \qquad (3.7.5.1)$$

where c is a closed curve around an infinitesimal area. Let us assume that the point $x(\tau_1)$ lies on the curve. Then we can parallel transport Z_μ around c using eq. 3.7.1.3. We approximate the affine connection with the first few terms of its power series expansion:

$$\Gamma^\sigma{}_{\lambda\mu}(x(\tau)) = \Gamma^\sigma{}_{\lambda\mu}(x(\tau_1)) + (x^a(\tau) - x^a(\tau_1))\, \partial\Gamma^\sigma{}_{\lambda\mu}(x(\tau_1))/\partial x^a + \ldots \qquad (3.7.5.2)$$

We expect the approximation to be good since c encloses an infinitesimal area. Substituting the expansion in eq. 3.7.1.3 we obtain an approximate expression for $Z_\mu(x(\tau))$:

$$Z_\mu(x(\tau)) \cong Z_\mu(x(\tau_1)) + (x^\lambda(\tau) - x^\lambda(\tau_1))\, \Gamma^\sigma{}_{\mu\lambda}(x(\tau_1))\, Z_\sigma(x(\tau_1)) \qquad (3.7.5.3)$$

up to first order in $x^\lambda(\tau) - x^\lambda(\tau_1)$. Combining eqs. 3.7.5.2 and 3.5.7.3, substituting in eq. 3.7.1.3, and forming an integral for we find $Z_\mu(x(\tau))$ to be:

$$Z_\mu(x(\tau)) \cong Z_\mu(x(\tau_1)) + \int_{\tau_1}^{\tau} d\tau \{\Gamma^\sigma{}_{\mu\rho}(x(\tau_1)) + (x^a(\tau) - {}^a(\tau_1))\partial\Gamma^\sigma{}_{\mu\rho}(x(\tau_1))/\partial x^a\} \bullet$$

$$\bullet\{Z_\sigma(x(\tau_1)) + (x^\lambda(\tau) - x^\lambda(\tau_1))\, \Gamma^\beta{}_{\sigma\lambda}(x(\tau_1))\, Z_\beta(x(\tau_1))\} dx^\rho/d\tau \quad (3.7.5.4)$$

to second order in $x^\lambda(\tau) - x^\lambda(\tau_1)$.

We now rewrite eq. 3.7.5.1 as an integral over a path parameterized by $x^\lambda(\tau)$:

$$I_c = \int_{\tau_1}^{\tau_2} d\tau\, Z_\mu(x)\, dx^\mu/d\tau \qquad (3.7.5.5)$$

where the value τ_2 will close the curve: $x^\lambda(\tau_2) = x^\lambda(\tau_1)$ for $\lambda = 0, 1, 2, 3$. Substituting eq. 3.7.5.4 in eq. 3.7.5.5 gives

$$I_c \cong \{\Gamma^\sigma{}_{\mu\rho}(x(\tau_1))\Gamma^\beta{}_{\sigma a}(x(\tau_1)) + \partial\Gamma^\beta{}_{\mu\rho}(x(\tau_1))/\partial x^a\} Z_\beta(x(\tau_1))\, J^{\mu a\rho}{}_c \quad (3.7.5.6)$$

to second order in $x^\lambda(\tau) - x^\lambda(\tau_1)$ after discarding some terms that become zero for the closed curve, The factor $J^{\mu a\rho}{}_c$ is given by

$$J^{\mu a\rho}{}_c = \int_{\tau_1}^{\tau_2} d\tau'\, dx^\mu/d\tau' \int_{\tau_1}^{\tau'} d\tau (x^a(\tau) - x^a(\tau_1))\, dx^\rho/d\tau \qquad (3.7.5.7)$$

which becomes

$$J^{\mu a\rho}{}_c = \int_{\tau_1}^{\tau_2} d\tau\, dx^\rho/d\tau [x^a(\tau) - x^a(\tau_1)][x^\mu(\tau_2) - x^\mu(\tau)] \qquad (3.7.5.8)$$

after switching the order of integration.

After some further simple algebra we find $J^{\mu a\rho}{}_c$ has an anti-symmetric term *and a symmetric term* in the indices a and ρ:

$$J^{\mu a\rho}{}_{c} = {}_{A}J^{\mu a\rho}{}_{c} + {}_{S}J^{\mu a\rho}{}_{c} \qquad (3.7.5.9)$$

where

$${}_{A}J^{\mu a\rho}{}_{c} = x^{\mu}(\tau_2)\int_{\tau_1}^{\tau_2} d\tau\, dx^{\rho}/d\tau\, x^{a}(\tau) + ½\int_{\tau_1}^{\tau_2} d\tau\, x^{\mu}[dx^{\rho}/d\tau\, x^{a}(\tau) - dx^{a}/d\tau\, x^{\rho}(\tau)] +$$

$$+ ½\int_{\tau_1}^{\tau_2} d\tau\, x^{\mu}(\tau)[x^{a}(\tau_1)\, dx^{\rho}/d\tau - x^{\rho}(\tau_1)\, dx^{a}/d\tau] \qquad (3.7.5.10)$$

and

$${}_{S}J^{\mu a\rho}{}_{c} = -½ \int_{\tau_1}^{\tau_2} d\tau\, x^{\rho}\, x^{a}\, dx^{\mu}/d\tau + ½\int_{\tau_1}^{\tau_2} d\tau\, x^{\mu}(\tau)[x^{a}(\tau_1)dx^{\rho}/d\tau + x^{\rho}(\tau_1)dx^{a}/d\tau] \qquad (3.7.5.11)$$

Eq. 3.7.5.6 can now be written

$$I_c \cong ½[R^{\beta}{}_{\mu\rho a}\, {}_{A}J^{\mu a\rho}{}_{c} + S^{\beta}{}_{\mu\rho a}\, {}_{S}J^{\mu a\rho}{}_{c}]\, Z_{\beta}(x(\tau_1)) \qquad (3.7.5.12)$$

where $R^{\beta}{}_{\mu\rho a}$ is the Riemann-Christoffel curvature tensor (eq. 3.5.4.1) and $S^{\beta}{}_{\mu\rho a}$ is a non-tensor which we will call the *symmetric curvature*:

$$S^{\beta}{}_{\mu\rho a} = \partial\Gamma^{\beta}{}_{\mu\rho}/\partial x^{a} + \partial\Gamma^{\beta}{}_{\mu a}/\partial x^{\rho} + \Gamma^{\sigma}{}_{\mu\rho}\Gamma^{\beta}{}_{\sigma a} + \Gamma^{\sigma}{}_{\mu a}\Gamma^{\beta}{}_{\sigma\rho} \qquad (3.7.5.13)$$

because it is symmetric in a and ρ. It does *not* have the index symmetries of $R^{\beta}{}_{\mu\rho a}$ (eqs. 3.5.4.6 – 3.5.4.8).

From eq. 3.7.5.12 we see that the line integral around a closed curve in several coordinates in complex space-time is not zero unless the curvature tensor and the symmetric curvature are both zero. In a flat space-time both are zero if the metric tensor is constant such as in the case of cartesian coordinates or light cone coordinates. If the metric tensor $g_{\mu\rho}$ is not constant such as in the case of a spherical spatial coordinate system, then $S^{\beta}{}_{\mu\rho a}$ is non-zero and line integrals around closed curves are non-zero in general. In a curved space-time both $R^{\beta}{}_{\mu\rho\alpha}$ and $S^{\beta}{}_{\mu\rho\alpha}$ are non-zero in general.

3.7.5.1 Parallel Transport of a Vector around a Closed Curve in Several Variables

The vector $Z_\mu(x(\tau))$ itself, given in eq. 3.7.5.4, when parallel transported around an infinitesimal closed curve at $x(\tau_1)$ will be unchanged if and only if the Riemann-Christoffel tensor $R^{\beta}{}_{\mu\rho\alpha}$ is zero at $x(\tau_1)$. Starting from eq. 3.7.5.4 it is easy to show that the net change is

$$\Delta Z_\beta = Z_\beta(x(\tau_2)) - Z_\beta(x(\tau_1)) \cong \tfrac{1}{2}R^{\alpha}{}_{\beta\rho\sigma}(\tau_1)Z_\alpha(x(\tau_1)) \oint d\tau\, x^\sigma dx^\rho/d\tau \quad (3.7.5.1.1)$$

Thus the change in the vector does not depend on the symmetric curvature $S^{\beta}{}_{\mu\rho\alpha}$.

3.7.5.2 Generalization of Cauchy's Theorem to Curved, Complex Space-time

In section 3.7.3 we considered line integrals in the complex plane of one coordinate and concluded that a line integral of a function around a closed curve in the complex plane of a single coordinate was zero if the curvature is zero and the function is holomorphic in the coordinate on, and inside, the curve (eq. 3.7.3.2). Eq. 3.7.5.12 enables us to make this statement more precise.

Suppose we consider the case of a vector function $Z_\mu(x(\tau))$ integrated around an infinitesimal closed curve c solely in the complex plane of the coordinate $x^{\mu 0}$ as developed from eq. 3.7.5.5. We assume the vector $Z_\mu(x)$ is holomorphic on c and on the surface enclosed by c. Then eq. 3.7.5.5 becomes

$$I_c = \int_{\tau_1}^{\tau_2} d\tau\, Z_{\mu_0}(x)\, dx^{\mu_0}/d\tau \quad (3.7.5.2.1)$$

with no implied sum over μ_0. Following the same line of development as section 3.7.5 we arrive at a special case of eq. 3.7.5.12 using the analyticity of Z_μ, and of the affine connection, to obtain the equivalent of eq. 3.7.5.4 and then to obtain

$$I_c \cong \tfrac{1}{2}[R^{\beta}{}_{\mu_0 \rho a}\,{}_A J^{\mu_0 a\rho}{}_c + S^{\beta}{}_{\mu_0 \rho a}\,{}_S J^{\mu_0 a\rho}{}_c]\, Z_\beta(x(\tau_1)) \qquad (3.7.5.2.2)$$

The line integral of an holomorphic vector around a closed infinitesimal curve in the complex plane of a single coordinate depends on the Riemann-Christoffel curvature tensor and the symmetric curvature. It is zero in a flat complex space-time. Thus we obtain a generalization of Cauchy's Theorem in curved complex space-time:

3.7.5.2.1 One Complex Variable Cauchy Theorem in Curved Complex Space-time

$$\int_c dx^{\mu_0} Z_{\mu_0}(x) = \tfrac{1}{2}[R^{\beta}{}_{\mu_0 \rho a}\,{}_A J^{\mu_0 a\rho}{}_c + S^{\beta}{}_{\mu_0 \rho a}\,{}_S J^{\mu_0 a\rho}{}_c]\, Z_\beta(x(\tau_1)) \qquad (3.7.5.2.1.1)$$

with no sum over μ_0.

3.7.5.2.2 Several Complex Variables Cauchy Theorem in Curved Complex Space-time

We also obtain the several complex variables generalization:

$$\int_c dx^{\mu} Z_{\mu}(x) = \tfrac{1}{2}[R^{\beta}{}_{\mu \rho a}\,{}_A J^{\mu a\rho}{}_c + S^{\beta}{}_{\mu \rho a}\,{}_S J^{\mu a\rho}{}_c]\, Z_\beta(x(\tau_1)) \qquad (3.7.5.2.2.1)$$

with an implied sum over μ.

3.7.5.3 Parallel Transport of a Vector Around a Closed Curve in the Complex Plane of a Coordinate

In section 3.7.5.3 we examined the line integral of a vector around a closed curve in the complex plane of one coordinate $x^{\mu 0}$. In this section we note that the parallel transport of a vector around a closed curve in the complex plane of a coordinate may be obtained directly from eq. 3.7.5.4. After some algebra we find

$$\Delta Z_\beta = Z_\beta(x(\tau_2)) - Z_\beta(x(\tau_1)) \cong [\partial\Gamma^{a}{}_{\beta\mu_0}/\partial x^{\mu_0} + \Gamma^{\nu}{}_{\beta\mu_0}\Gamma^{a}{}_{\nu\mu_0}] Z_a(x(\tau_1)) \oint x^{\mu_0} dx^{\mu_0}$$
$$= 0 \qquad (3.7.5.3.1)$$

(no sum over the μ_0 index) due to the integral around the closed path being zero. Eq. 3.7.5.3.1 is based on expansions (eqs. 3.7.5.2 and 3.7.5.3) that require analyticity in Z_μ and the affine connection.

3.7.5.4 Path Independence of Parallel Transported Vectors in Complex Space-time

In section 3.7.5.1 we examined the parallel transport of a vector around a closed infinitesimal curve and found that the change in the vector was zero if the Riemann-Christoffel curvature tensor was zero in the neighborhood of the closed curve (eq. 3.7.5.1.1). By combining closed infinitesimal curves in a grid pattern one can construct a finite curve for which the conclusion also applies; namely, *that the parallel transport of a vector around a closed finite curve leaves the vector unchanged if the Riemann-Christoffel curvature tensor is zero in a domain D containing the curve.*

Therefore the parallel transport of a vector between two points in D, x_1 and x_2, is independent of the choice of curve (within D) between x_1 and x_2. The components of the vector at x_2 are the same independent of the curve (within D) along which the vector is transported *if the Riemann-Christoffel curvature tensor is zero in D.*

3.7.5.5 Path Independence of Line Integrals of Vectors in Complex Space-time

In section 3.7.5 we examined the line integral of a vector around a closed infinitesimal curve and found that the value of the line integral was zero if the Riemann-Christoffel curvature tensor and the symmetric curvature were both zero in the neighborhood of the closed curve (eq. 3.7.5.12).

Again we can combine closed infinitesimal curves in a grid pattern to construct a line integral around a finite curve for which the conclusion also applies; namely, *that the line integral of a vector around a closed finite curve is zero if the Riemann-Christoffel curvature tensor and the symmetric curvature are both zero in a domain D containing the curve (eq. 3.7.5.2.2.1).*

Therefore the value of the line integral of a vector between two points in D, x_1 and x_2, is independent of the choice of curve (within D) between x_1 and x_2. *if the Riemann-Christoffel curvature tensor and the symmetric curvature are both zero in D.*

Since the symmetric curvature is not a tensor, a change of coordinate system can transform it from a zero to a non-zero value or vice versa. If the

Riemann-Christoffel curvature tensor is zero in one coordinate system it remains zero when a transformation to another coordinate system is made.

4.5 On 8-Dimensional Complex Coordinate Transformations

We can use a matrix notation to establish the form of complex coordinate transformations. We start by defining 4-dimensional complex coordinate transformations in matrix form. Then we define a set of 8-dimensional complex coordinate transformations. Our set of transformations is a subset of the complete set of 8-dimensional complex coordinate transformations.

4.5.1 Matrix Representation of 4 Complex Dimensions Coordinate Transformations

Let us begin by considering a local transformation in 4-dimensional complex space-time that transforms the flat space metric tensor $\eta_{\alpha\beta}$ where x^μ represents the flat space coordinates to a new form $g_{\mu\nu}$ in the coordinate system x'^μ

$$g_{\mu\nu}(x') = \eta_{\alpha\beta}\,\partial x^\alpha/\partial x'^\nu \; \partial x^\beta/\partial x'^\nu \tag{4.5.1.1}$$

This can be expressed in matrix form as

$$\left(g_{\mu\nu}(x')\right) = \Lambda^T\left(\partial x^\alpha/\partial x'^\nu\right)\left(\eta_{\alpha\beta}\right)\Lambda\left(\partial x^\beta/\partial x'^\nu\right) \tag{4.5.1.2}$$

where Λ is a complex (generally) matrix formed of partial derivatives, and Λ^T is its transpose. We will write eq. 4.5.1.2 in the abbreviated form:

$$(g) = (\Lambda^T)(\eta)(\Lambda) \tag{4.5.1.3}$$

4.5.2 Matrix Representation of 8 Complex Dimensions Coordinate Transformations

We now use the matrices defined in the preceding section to define an 8-dimensional matrix formulation. We define an 8-dimensional Minkowski metric matrix with

$$(\eta_8) = \begin{bmatrix} (\eta) & (0) \\ (0) & (\eta) \end{bmatrix} \qquad (4.5.2.1)$$

with the usual 4 by 4 Minkowski metric matrices along the diagonal. We then define the 8-dimensional transformation matrix in terms of the 4-dimensional transformation matrix Λ.

$$(\Lambda_8) = \begin{bmatrix} (\Lambda) & (0) \\ (0) & (\Lambda^*) \end{bmatrix}$$

(4.5.2.2)

and the 8-dimensional metric $g_{8\mu\nu}$ with

$$(g_8) = \begin{bmatrix} (g) & (0) \\ (0) & (g^*) \end{bmatrix}$$

(4.5.2.3)

We then can write the 8-dimensional version of eq. 4.5.3:

$$(g_8) = (\Lambda_8^{\,T})(\eta_8)\,(\Lambda_8) \qquad (4.5.2.4)$$

which implies eq. 4.5.1.3 and its complex conjugate:

$$(g^*) = (\Lambda^{T*})(\eta)\,(\Lambda^*) \qquad (4.5.2.5)$$

since $\eta = \eta^*$.

4.6 The Complex Space-time Determinant g

In this section we evaluate the complex space-time determinant g. First we note:

$$I = \int dx^0 dx^{0*} dx^1 dx^{1*} dx^2 dx^{2*} dx^3 dx^{3*} f(x^0, x^{0*}, x^1, x^{1*}, x^2, x^{2*}, x^3, x^{3*})$$

$$\equiv \int d^4x d^4x^* \; f(x,x^*) = \int d^4x' d^4x'^* \partial(x,x^*)/\partial(x',x'^*) \; f(x,x^*)$$

where

$$\partial(x,x^*)/\partial(x',x'^*) = \det \begin{bmatrix} (\partial x^\mu/\partial x'^\nu) & (0) \\ (0) & (\partial x^{\mu *}/\partial x'^{\nu *}) \end{bmatrix}$$

$$= \det(\Lambda_8) = \det(\Lambda)[\det(\Lambda)]^* \qquad (4.6.1)$$

The matrix equations in section 4.5 enable us to relate the determinant of the metric tensor to the local transformation relating $g_{\mu\nu}$ to $\eta_{\alpha\beta}$. Eq. 4.5.1.3 implies

$$\det(g) = \det(\Lambda^T)\det(\eta)\det(\Lambda) \qquad (4.6.1)$$

and thus gives

$$g = \det(g) = -[\det(\Lambda)]^2 \qquad (4.6.2)$$

using $\det(\Lambda^T) = \det(\Lambda)$ and $\det(\eta) = -1$. Thus

$$g = -[\det(\Lambda)]^2 = -[\partial(x^0, x^1, x^2, x^3)/\partial(x'^0, x'^1, x'^2, x'^3)]^2 \qquad (4.6.3)$$

is minus the square of a Jacobian just as in the case of real space-time. The complex conjugate g* is determined from the complex conjugate of eq. 4.6.3

$$g^* = -\left[\det(\Lambda)\right]^{2*} = -\left[\partial(x^{0*}, x^{1*}, x^{2*}, x^{3*})/\partial(x'^{0*}, x'^{1*}, x'^{2*}, x'^{3*})\right]^2 \quad (4.6.4)$$

If we take the determinant of eq. 4.5.2.4 we find

$$g_8 \equiv \det(g_8) = \left[\det(\Lambda_8)\right]^2 = \left[\det(\Lambda)\right]^2\left[\det(\Lambda)\right]^{2*} = gg^* \quad (4.6.5)$$

using eq. 4.6.1.

4.7 Volume Integrals using Complex Coordinates

Eq. 4.1.1 defines a volume integral in a real, 8-dimensional local Lorentz frame. Eq. 4.1.7 re-expresses that integral in terms of complex coordinates:

$$I = \int dx^0dx^{0*}dx^1dx^{1*}dx^2dx^{2*}dx^3dx^{3*}\ \partial(x,y)/\partial(x,x^*)$$
$$f(x^0,x^{0*},x^1,x^{1*},x^2,x^{2*},x^3,x^{3*}) \quad (4.1.7)$$

which becomes

$$I = \int dx^0dx^{0*}dx^1dx^{1*}dx^2dx^{2*}dx^3dx^{3*}f(x^0,x^{0*},x^1,x^{1*},x^2,x^{2*},x^3,x^{3*})/16 \quad (4.7.1)$$

in the local Lorentz frame using the inverse of eq. 4.2.6. If we transform to a different complex coordinate system then we find that the transformation leads to

$$dx^0dx^{0*}dx^1dx^{1*}dx^2dx^{2*}dx^3dx^{3*} = \partial(x,x^*)/\partial(x',x'^*)\ dx'^0dx'^{0*}dx'^1dx'^{1*}dx'^2dx'^{2*}dx'^3dx'^{3*}$$
$$(4.7.2)$$

with $\partial(x,x^*)/\partial(x',x'^*)$ defined by eq. 4.2.2. Using eqs. 4.6.1 and 4.6.6 in conjunction with eqs. 4.7.1 and 4.7.2 gives

$$I = (\tfrac{1}{2}\, i)^{-4}\int dx'^0dx'^{0*}dx'^1dx'^{1*}dx'^2dx'^{2*}dx'^3dx'^{3*}(gg^*)^{1/2}f(x') \quad (4.7.3)$$

which we abbreviate to

$$I = 2^{-4}\int d^4x'd^4x'^*(gg^*)^{1/2}f(x') \quad (4.7.4)$$

Thus the *proper volume element* in our complex space-time formulation is

$$2^{-4}d^4xd^4x^*\ (gg^*)^{\frac{1}{2}} \tag{4.7.5}$$

4.8 Gauss' Theorem for Complex Space-time

Gauss' Theorem plays an important role in the analysis of features of general relativity. In this section we will derive the complex space-time version of Gauss' Theorem. It is significantly different from the real space-time equivalent.

Consider the integral of the covariant divergence of a holomorphic vector function F of the *holomorphic coordinates* x^μ: $F^\kappa(x)$ over a volume V. (We call the complex conjugate of the x^μ coordinates *anti-holomorphic coordinates* and denote them as $x^{\mu *}$.)

$$I = 2^{-4}\int_V d^4xd^4x^*\ (gg^*)^{\frac{1}{2}}\ F^\rho{}_{;\rho} \tag{4.8.1}$$

Now

$$F^\rho{}_{;\rho} = g^{-\frac{1}{2}}\ \partial(g^{\frac{1}{2}}F^\rho)/\partial x^\rho \tag{4.8.2}$$

by eq. 3.5.3.8. Therefore eq. 4.8.1 becomes

$$I = 2^{-4}\int_V d^4xd^4x^*\ g^{*\frac{1}{2}}\ \partial(g^{\frac{1}{2}}F^\rho)/\partial x^\rho \tag{4.8.3}$$

Since the integral over the x variables is an exact divergence of F^ρ with respect to x^ρ we can use a variation of the complex form of Green's Theorem to eliminate one of the integrations over x thus producing a "surface" integral. The one complex variable Green's Theorem that we will use is:

Alternate One Complex Variable Version of Green's Theorem

If $F(z, z^*)$ is continuous and has continuous partial derivatives in some region R and on its boundary curve C then

$$\int_R dzdz^*\ \partial F/\partial z = \oint_C dz^*\ F(z, z^*) \tag{4.8.4}$$

Applying this theorem to each of the four terms in the integrand of eq. 4.8.3 we obtain:

$$I = 2^{-4}\int d\Sigma_\rho\,(gg^*)^{1/2}\,F^\rho \qquad (4.8.5)$$

where

$$\int_V d\Sigma_\rho \equiv \int_V \prod_{\mu\neq\rho} dx^\mu dx^{\mu *} \oint_{C_\rho} dx_\rho^* \qquad (4.8.6)$$

with C_ρ the curve bounding the area of the x_ρ integration and the other integrals spanning the volume of integration in those variables. Thus we have the

4.8.1 Complex Space-time Gauss' Theorem

$$\int_V d^4x d^4x^*(gg^*)^{1/2}F^\rho{}_{;\rho} = \int_V d\Sigma_\rho\,(gg^*)^{1/2}F^\rho \qquad (4.8.1.1)$$

In addition the right side of eq. 4.8.1.1 displays the form of a *proper surface integral*. The next section illustrates the application of the Complex Space-time Gauss' Theorem. The generalization to higher dimensional complex spaces is direct.

4.9 Examples of Gauss' Theorem in Complex Space-time

In this section we will consider some examples of the complex space-time Gauss' Theorem.

4.9.1 Four-dimensional, Flat, Complex Space-time Gauss' Theorem Example: a Hypercube

We will consider the volume integration of the divergence of a vector, $V^\rho(x)$, for the case of a 4-dimensional hypercube with sides of length a in flat space-time ($g = g^* = -1$). Let

$$V^\rho(x) = x^\rho \qquad (4.9.1.1)$$

The volume integral by direct integration of the real and imaginary parts is

$$I = \int dx^0 dy^0 dx^1 dy^1 dx^2 dy^2 dx^3 dy^3 \, (gg^*)^{1/2} V^{\rho}{}_{;\rho} = 4a^8 \qquad (4.9.1.2)$$

The integration of the left side of the complex space-time Gauss' Theorem (eq. 4.8.1.1) multiplied by 2^{-4} is

$$I = 2^{-4}\int d^4x d^4x^* \, (gg^*)^{1/2} \, V^{\rho}{}_{;\rho} = 2^{-4}\int d^4x d^4x^* 4 \qquad (4.9.1.3)$$

$$= 2^{-2}\left[\int dz dz^*\right]^4 = 2^{-2}(-2ia^2)^4 = 4a^8 \qquad (4.9.1.4)$$

The integration of the right side of the complex space-time Gauss' Theorem (eq. 4.8.1.1) multiplied by 2^{-4} is

$$J = 2^{-4}\int d\Sigma_\rho \, (gg^*)^{1/2} \, V^\rho = 2^{-4}\int \prod_{\mu \neq \rho} dx^\mu dx^{\mu *} \oint_{C_\rho} dx_\rho{}^* \, x^\rho \qquad (4.9.1.5)$$

$$J = 2^{-4}\int \sum_\rho \prod_{\mu \neq \rho} dx^\mu dx^{\mu *} (-2ia^2) \qquad (4.9.1.6)$$

using the complex variable result:

$$\oint dz^* \, z = -2iA \qquad (4.9.1.7)$$

where A is the area enclosed by the curve for each of the 4 terms in the sum in eq. 4.9.1.5. For each term $A = a^2$. Performing the remaining integrals in eq. 4.9.1.6 yields:

$$J = 2^{-4}(-2ia^2)4(-2ia^2)^3 = 4a^8 = I \qquad (4.9.1.8)$$

proving the theorem.

4.9.2 Two-dimensional, Flat, Complex Space-time Gauss' Theorem Example: a Complex Sphere

We will consider the volume integration of the divergence of a vector, $V^\rho(x)$, for the case of a sphere in a flat, 2-dimensional complex Euclidean space with radius r. $(g = g^* = -1)$. Let

$$V^\rho(x) = x^\rho \tag{4.9.2.1}$$

again. The volume integral by direct integration of the real and imaginary parts is

$$I = \int dx^1 dy^1 dx^2 dy^2 \, (gg^*)^{1/2} V^\rho{}_{;\rho} = \pi^2 r^4 \tag{4.9.2.2}$$

The integration of the left side of the complex space-time Gauss' Theorem (eq. 4.8.1.1) multiplied by $(i/2)^2$ is

$$I = -2^{-2}\int d^2x d^2x^* \, (gg^*)^{1/2} \, V^\rho{}_{;\rho} = -2^{-2}\int d^2x d^2x^* 2 \tag{4.9.2.3}$$

$$= -2^{-1}\int dx^1 dx^{1*} \, (\text{-}2i)\pi(r^2 - x^1 x^{1*}) \tag{4.9.2.4}$$

Using the alternate form of Green's Theorem (eq. 4.8.4) eq. 4.9.2.4 becomes

$$I = i\pi \oint_C dx^{1*} \, [r^2 x^1 - (x^1)^2 x^{1*}] = i\pi[-2i\pi r^4 + i\pi r^4] = \pi^2 r^4 \tag{4.9.2.5}$$

The integration of the right side of the complex space-time Gauss' Theorem (eq. 4.8.1.1) multiplied by $(i/2)^{-2}$ is

$$J = -2^{-2}\int d\Sigma_\rho \, (gg^*)^{1/2} \, V^\rho = -2^{-2}\Big[\int dx^2 dx^{2*} \oint_C dx^{1*} \, x^1 + dx^1 dx^{1*} \oint_C dx^{2*} \, x^2\Big] \tag{4.9.2.6}$$

$$= -2^{-2}2(-2i)\int dx dx^* \, \pi(r^2 - xx^*) = \pi^2 r^4 \tag{4.9.2.7}$$

using eq. 4.9.1.7 and following steps similar to eq. 4.9.2.5 thus confirming the theorem.

4.10 Some Other Useful Relations for Complex Variable Integrations

Some other relations that are helpful in evaluating integrals over complex variables are:

4.10.1 One Complex Variable Version of Green's Theorem

If $F(z, z^*)$ is continuous and has continuous partial derivatives in some region R and on its boundary curve C then

$$\int_R dz dz^* \, \partial F/\partial z^* = -\oint_C dz \, F(z, z^*) \qquad (4.10.1.1)$$

4.10.2 An Area Integral

The following is a useful integral: the integral around a closed curve C is

$$\oint_C dz \, z^* = 2iA \qquad (4.10.2.1)$$

where A is the area enclosed by the curve C.

REFERENCES

Blaha, S., 1998, *Cosmos and Consciousness* (Pingree-Hill Publishing, Auburn, NH, 1998).

_______2003, *A Finite Unified Quantum Field Theory of the Elementary Particle Standard Model and Quantum Gravity Based on New Quantum Dimensions™& a New Paradigm in the Calculus of Variations* (Pingree-Hill Publishing, Auburn, NH, 2003).

_______2004, *Quantum Big Bang Cosmology: Complex Space-time General Relativity, Quantum Coordinates™ Dodecahedral Universe, Inflation, and New Spin 0, ½, 1 & 2 Tachyons & Imagyons* (Pingree-Hill Publishing, Auburn, NH, 2004).

______ *2005a, Quantum Theory of the Third Kind: A New Type of Divergence-free Quantum Field Theory Supporting a Unified Standard Model of Elementary Particles and Quantum Gravity based on a New Method in the Calculus of Variations* (Pingree-Hill Publishing, Auburn, NH, 2005).

______, 2005b, *The Metatheory of Physics Theories, and the Theory of Everything as a Quantum Computer Language* (Pingree-Hill Publishing, Auburn, NH, 2005).

______, 2005c, *The Equivalence of Elementary Particle Theories and Computer Languages: Quantum Computers, Turing Machines, Standard Model, Superstring Theory, and a Proof that Gödel's Theorem Implies Nature Must Be Quantum* (Pingree-Hill Publishing, Auburn, NH, 2005).

______, 2006, *A Derivation of ElectroWeak Theory based on an Extension of Special Relativity; Black Hole Tachyons; & Tachyons of Any Spin.* (Pingree-Hill Publishing, Auburn, NH, 2006).

______, 2007a, *Physics Beyond the Light Barrier: The Source of Parity Violation, Tachyons, and A Derivation of Standard Model Features* (Pingree-Hill Publishing, Publishing, Auburn, NH, 2007).

______, 2007b, *The Origin of the Standard Model: The Genesis of Four Quark and Lepton Species, Parity Violation, the ElectroWeak Sector, Color SU(3), Three Visible Generations of Fermions, and One Generation of Dark Matter with Dark Energy* (Pingree-Hill Publishing, Auburn, NH, 2007).

______, *2008a, A Direct Derivation of the Form of the Standard Model From GL(16) (Pingree-Hill Publishing, Auburn, NH, 2008).*

______, 2008b, *A Complete Derivation of the Form of the Standard Model With a New Method to Generate Particle Masses Second Edition* (Pingree-Hill Publishing, Auburn, NH, 2008)

______, 2009, *The Algebra of Thought & Reality: The Mathematical Basis for Plato's Theory of Ideas, and Reality Extended to Include A Priori Observers and Space-Time Second Edition* (Pingree-Hill Publishing, Auburn, NH, 2009).

______, 2010a, *Operator Metaphysics: A New Metaphysics Based on a New Operator Logic and a New Quantum Operator Logic that Lead to a Mathematical Basis for Plato's Theory of Ideas and Reality* (Pingree-Hill Publishing, Auburn, NH, 2010).

______, 2010b, *The Standard Model's Form Derived from Operator Logic, Superluminal Transformations and GL(16)* (Pingree-Hill Publishing, Auburn, NH, 2010).

______, 2011a, *21st Century Natural Philosophy Of Ultimate Physical Reality* (McMann-Fisher Publishing, Auburn, NH, 2011).

______, 2011b, *All the Universe! Faster Than Light Tachyon Quark Starships & Particle Accelerators with the LHC as a Prototype Starship Drive Scientific Edition* (Pingree-Hill Publishing, Auburn, NH, 2011).

______, 2011c, *From Asynchronous Logic to The Standard Model to Superflight to the Stars* (Blaha Research, Auburn, NH, 2011).

______, 2012a, *From Asynchronous Logic to The Standard Model to Superflight to the Stars volume 2: Superluminal CP and CPT, U(4) Complex General Relativity and The Standard Model, Complex Vierbein General Relativity, Kinetic Theory, Thermodynamics* (Blaha Research, Auburn, NH, 2012).

______, 2012b, *Standard Model Symmetries, And Four And Sixteen Dimension Complex Relativity; The Origin Of Higgs Mass Terms* (Blaha Reasearch, Auburn, NH, 2012).

______, 2013, *The Bridge to Dark Matter; A New Sister Universe; Dark Energy; Inflatons; Quantum Big Bang; Superluminal Physics; An Extended Standard Model Based on Geometry* (Blaha Reasearch, Auburn, NH, 2013).

Eddington, A. S., 1952, *The Mathematical Theory of Relativity* (Cambridge University Press, Cambridge, U.K., 1952).

Evelyn-White, H. G. (tr), 1914, *Hesiod, The Homeric Hymns and Homerica* (Loeb Classical Library, Harvard University Press, Cambridge, MA, 1914).

Fant, Karl M., 2005, *Logically Determined Design: Clockless System Design With NULL Convention Logic* (John Wiley and Sons, Hoboken, NJ, 2005).
Gradshteyn, I. S. and Ryzhik, I. M., 1965, *Table of Integrals, Series, and Products* (Academic Press, New York, 1965).
Rescher, N., 1967, *The Philosophy of Leibniz* (Prentice-Hall, Englewood Cliffs, NJ, 1967).
Sakurai, J. J., 1964, *Invariance Principles and Elementary Particles* (Princeton University Press, Princeton, NJ, 1964).
Streater, R. F. and Wightman, A. S., 2000, *PCT, Spin, Statistics, and All That* (Princeton University Press, Princeton, NJ 2000).
Weinberg, S., 1995, *The Quantum Theory of Fields Volume I* (Cambridge University Press, New York, 1995).
Weyl, H., 1950, *Space, Time, Matter* (Dover, New York, 1950).
Weyl, H., (Tr. S. Pollard et al), 1987, *The Continuum* (Dover Publications, New York, 1987).

INDEX

About the Author

Stephen Blaha is an internationally known physicist with interests in Science, the Arts, and Technology. He had an Alfred P. Sloan Foundation scholarship in college. He received his Ph.D. in Physics from Rockefeller University. He has served on the faculties of several major universities. He was also a Member of the Technical Staff at Bell Laboratories, a manager at the Boston Globe Newspaper, a Director at Wang Laboratories, and President of Blaha Software Inc and of Janus Associates Inc. (NH).

Among other achievements he was a co-discoverer of the "r potential" for heavy quark binding developing the first (and still the only demonstrable) non-abelian gauge theory with an "r" potential; first suggested the existence of topological structures in superfluid He-3; first proposed Yang-Mills theories would appear in condensed matter phenomena with non-scalar order parameters; first developed a grammar-based formalism for quantum computers and applied it to elementary particle theories; first developed a new form of quantum field theory without divergences (thus solving a major 60 year old problem that enabled a unified theory of the Standard Model and Quantum Gravity without divergences to be developed); first developed a formulation of complex General Relativity based on analytic continuation from real space-time; first developed a generalized non-homogeneous Robertson-Walker metric that enabled a quantum theory of the Big Bang to be developed without singularities at $t = 0$; first generalized Cauchy's theorem and Gauss' theorem to complex, curved multi-dimensional spaces; received Honorable Mention in the Gravity Research Foundation Essay Competition in 1978; first developed a physically acceptable theory of faster-than-light particles; first showed a universe with three complex spatial dimensions is icosahedral; first derived a composition of extrema method in the Calculus of Variations; first quantitatively suggested that inflationary periods in the history of the universe were not needed; first proved Gödel's Theorem implies Nature must be quantum; provided a new alternative to the Higgs Mechanism, and Higgs particles, to generate masses; first showed how to resolve logical paradoxes including Gödel's Undecidability Theorem by developing Operator Logic and Quantum Operator Logic; first developed a quantitative harmonic oscillator-like model of the life cycle, and interactions, of civilizations; first showed how equations describing superorganisms also apply to civilizations; and first developed an axiomatic derivation of the forms of The Standard Model with DARK PARTICLEs from geometry – space-time properties – The faster than light Standard Model.

He has had a major impact on a succession of elementary particle theories: his Ph.D. thesis (1970), and papers, showed that quantum field theory calculations to all orders in ladder approximations could not give scaling deep inelastic electron-nucleon scattering. He later showed the eigenvalue equation for the fine structure constant α in Johnson-Baker-Willey QED had a zero at $\alpha = 1$ not 1/137 by solving the Schwinger-Dyson equations to all orders in an approximation that agreed with exact results to 8th order in α thus ending interest in this theory. In 1979 at Prof. Ken Johnson's (MIT) suggestion he calculated the proton-neutron mass difference in the MIT bag model and found the result had the wrong sign reducing interest in the bag model. These results all appear in Physical Review papers. In the 2000's he repeatedly pointed out the shortcomings of SuperString theory and showed that The Standard Model's form could be derived from space-time geometry by an extension of Lorentz transformations to faster than light transformations. This deeper space-time basis greatly increases the possibility that it is part of THE fundamental theory.

In the early 1980's Blaha was also a pioneer in the development of UNIX for financial, scientific and Internet applications: benchmarked UNIX versions showing that block size was critical for UNIX performance, developing financial modeling software, starting database benchmarking comparison studies, developing Internet-like UNIX networking (1982) and developing a hybrid shell programming technique (1982) that was a precursor to the PERL programming language. He was also the manager of the AT&T ten-year future products development database. His work helped lead to commercial UNIX on computers such as Sun Micros, IBM AIX minis, and Apple computers.

In the 1980's he pioneered the development of PC Desktop Publishing on laser printers. and was nominated for three "Awards for Technical Excellence" in 1987 by PC Magazine for PC software products that he designed and developed.

In the past ten years Dr. Blaha has written over 35 books on a wide range of topics. Some recent major works are: *From Asynchronous Logic to The Standard Model to Superflight to the Stars, All the Universe!* and *SuperCivilizations: Civilizations as Superorganisms.*

www.ingramcontent.com/pod-product-compliance
Lightning Source LLC
Chambersburg PA
CBHW081535220326
41598CB00036B/6439
* 9 7 8 0 9 8 9 3 8 2 6 2 5 *